United Nations University Series on Regionalism

Volume 27

This book series offers a platform for innovative work on (supra-national) regionalism from a global and inter-disciplinary perspective. It started with the World Reports on Regional Integration, published in collaboration with other UN agencies, and now publishes theoretical, methodological and empirical contributions related to various themes on interregionalism from academics and policy-makers worldwide. All book proposals are reviewed by an International Editorial Board and final approval on manuscripts also comes from the board.

The series editors are particularly interested in book proposals dealing with:

- comparative regionalism;
- comparative work on regional organizations;
- interregionalism
- the role of regions in a multi-level governance context;
- the interactions between the UN and the regions;
- the regional dimensions of the reform processes of multilateral institutions;
- the dynamics of cross-border micro-regions and their interactions with supra-national regions;
- methodological issues in regionalism studies.

Accepted book proposals can receive editorial support from UNU-CRIS for the preparation of manuscripts.

Please send book proposals to: pdelombaerde@cris.unu.edu and lvanlangenhove@cris.unu.edu

Andrea Ribeiro Hoffmann • Paula Sandrin
Yannis E. Doukas

Editors

Climate Change in Regional Perspective

European Union and Latin American
Initiatives, Challenges, and Solutions

Springer

Editors
Andrea Ribeiro Hoffmann
Institute of International Relations
Pontifical Catholic University
of Rio de Janeiro
Gávea, Rio de Janeiro, Brazil

Paula Sandrin
Institute of International Relations
Pontifical Catholic University
of Rio de Janeiro
Gávea, Rio de Janeiro, Brazil

Yannis E. Doukas
Agricultural Development, Agri-food
and Natural Resources Management
National and Kapodistrian University
of Athens
Athens, Greece

ISSN 2214-9848   ISSN 2214-9856   (electronic)
United Nations University Series on Regionalism
ISBN 978-3-031-49328-7   ISBN 978-3-031-49329-4   (eBook)
https://doi.org/10.1007/978-3-031-49329-4

This Springer imprint is published by the registered company Springer Nature Switzerland AG
The registered company address is: Gewerbestrasse 11, 6330 Cham, Switzerland

Paper in this product is recyclable.

Foreword

The Jean Monnet Network "Crisis-Equity-Democracy for Europe and Latin America" (JMN) was established in 2015 to research and exchange on crises, their management, and their economic, social and political impacts in Europe and Latin America. The approach of the network included a comparative regionalism perspective, looking, on the one hand, at the role and impact of regional governance in and on crises, their management and resilience, and, on the other hand, on the consequences of crises for regionalism and integration. Last but not least, the JMN strives to elaborate proposals to promote bi-lateral cooperation between the regions in regard to crisis management and with the purpose to increase resilience and their impact in global governance.

The JMN's initial point of departure was the financial crisis of 2007/8, the following sovereign debt crisis with the austerity programmes provoking a wave of public outrage, the mobilization of social movements and protests globally, the loss of trust in the established political parties and institutions and as a consequence the mushrooming of new political parties and the polarization of politics.

Along the lifespan and debate of the first term of the Jean Monnet Network, it became increasingly clear that we are currently facing not only an economic governance crisis, but rather a global governance crisis with the multilateral governance system increasingly being put into question, a democratic crisis or crisis of democracy with the rise of autocrats and populism and, last but not least, a climate crisis, which came increasingly to the forefront of the debate and "naturally" became the focus of the second mandate of the Jean Monnet Network. This joint volume, which reflects critically about ongoing agendas on the environment and climate change in Europe and Latin America, aims to contribute to the dialogue between the European Union (EU) and the Community of Latin American and Caribbean States (CELAC) by advancing recommendations on how to improve the mechanisms to address these issues in the context of the bi-regional strategic partnership.

I want to thank all the members of the network for their commitment and excellent work. Particular thanks to Andrea Ribeiro Hoffmann for taking over the leadership in the second term of the JMN from the original Coordinators, Christian Ghymers and myself, and to Christian for his continuous engagement and efforts to promote the network.

While financial and/or economic crises are a frequent and recurrent matter, climate change and global warming are a question of survival and need to be addressed urgently and with an all- encompassing approach. There is no plan(et B). Environmental degradation and the climate crisis are indeed global matters. All over the world, people suffer from an increasing number of heat waves, draughts, fires, flooding, mudslides, vanishing islands and other impacts of global warming.

Environmental degradation can be felt and became visible to everybody leading to general awareness on the matter with protest and social or rather environmental mobilization, green parties increasingly gaining ground in elections and joining government coalitions. As a logical consequence, green agendas have been taken over by most of the political parties into their party programmes and later into government programmes, with the exception of mainly right-wing populist parties and some reluctance by more traditional industry-friendly ones. This "green revolution" also captured the European Union. After the last elections of the European parliament in 2019, with record participation reinforcing its legitimacy, the EU adopted the "Green Deal" as its key programme to engage with the green transition both in internal policies and in its foreign policies. Green policies are key for investments in the Next Generation EU and the Recovery and Resilience Facility (RRF) with the objective to help the EU achieve its target of climate neutrality by 2050 and to grow out of the crisis caused by the Covid-19 pandemic and to increase energy independence and strategic autonomy within the RepowerEU programme as an answer to the War in Ukraine.

Green policies have become an integral part of economic governance and green objectives have also entered financial markets with the guidance provided by the EU's "taxonomy", the European Central Bank's guidance and supervision and the leading role of the European Investment Bank in leveraging green investment as the EU's Climate Bank as described in this volume by Stephany Griffith-Jones. Environmental policies have grown from being an objective among others to become the key objective per se, an integral and essential instrument to promote growth, increase resilience and improve strategic autonomy.

Yet all these measures seem to be too little and too late. UN Secretary-General Antonio Guterres warned on April 20, 2023, addressing the fourth Major Economies Forum on Energy and Climate (MEF) convened by US President that "Today's policies would make our world 2.8 degrees hotter by the end of the century. And this is a death sentence."

Andrea Ribeiro Hoffmann, Paula Sandrin and Yannis E. Doukas point out in the conclusion of this volume the discrepancy of the EU and Latin American regional organizations with the latter falling short of including environmental concerns and climate change in their priorities, in spite of the fact that Latin American countries have increasingly addressed these concerns in their domestic policies and in their participation in global multilateral institutions over the last years. The lack of epistemic communities and the strong lobby of agrobusiness sectors in most LAC countries, including Brazil, contribute to this matter, as Andrea Ribeiro Hoffmann argues in her chapter. She recommends that LAC regional organizations foment debates and include commitments to address environmental and climate change problems as a priority.

Notwithstanding the existential risks of the climate change crisis, there are forces in society which are in climate denialism and as Paula Sandrin elaborates in her chapter in this volume, we are also confronted with climate disavowal, when climate change is acknowledged, but ineffective responses to mitigate it keep being repeated.

Additionally, there are business and industry lobbies and political circles undermining the efforts by "greenwashing" or even by committing fraud as was the case of the Volkswagen emission scandal, to give an example from the industry sector. There is slowing down or watering down the process or negotiating exceptions at political level as is the case of the watering down of the ban of new cars with combustion engines or declaring nuclear energy as a renewable energy.

The strongest resistance to a paradigm change towards a committed all-encompassing programme to combat global warming can be found in extreme right wing populist parties, which however could have major consequence for the environment as well as for multilateral and global governance, as we could witness in the Americas over the past years, with the biggest countries both in North and in South America entering a "dark age" under the spell of populist leaders. Unfortunately, the damage caused by this era in regard to environmental deterioration as well as in terms of multilateral governance might not be reversible. Time is of the essence in the fight against climate change. UN Secretary-General Antonio Guterres emphasized at the MEF that the possibility of limiting global temperature rise to 1.5 °C requires a "quantum leap in climate action" (…) "We need global acceleration through cooperation. And that means rising above disagreements, differences and tensions."[1] Unfortunately, we have lost years to counteract global warming with governments in the lead of two major American nations, USA and Brazil, boycotting the global alliance on climate change, hazardously furthering the destruction of the environment and neglecting the protection of the Amazon Forest, the lungs of the Earth, just to mention an obvious example.

While the two populist leaders, Donald Trump and Jair Bolsonaro, have not succeeded to be re-elected, their successors are confronted with the difficult task to re-build "trust" in democracy, to counteract polarization or even "pacify" the polarized population in their country. At the same time, they need to re-gain trust, credibility and the reputation of their country's commitments to combat climate change and their respective roles in regional governance and global governance and in the multilateral governance system which their countries had defied for the past years.

---

[1] UN News, Global perspective human stories, 20 April 2023, https://news.un.org/en/story/2023/04/1135862.

However, the putting into question of the multilateral governance system such as the World Trade Organization in the case of the USA and the neglect of the regional leadership role in the case of Brazil left scars and provoked reactions. Other international players have filled the gaps and increased their power and role promoting different rules and value or expanding their influence by offering cooperation without the conditionality of democracy, human rights and rule of law.

On the one hand this development leads to the positive impact that there is now more interest for certain developing countries, particularly in Africa, and even a "competition for cooperation and partnerships" with them, on the other hand, these new partnerships facilitate the rise of autocratic regimes violating democratic principles and human rights.

Interestingly, "competition" has turned into a key word in regard to the environment. While the USA and Brazil went into the reverse direction during their climate denialist governments, China and Europe initiated a race to win leadership in green and renewable technology, which increasingly turned key for future competitiveness. Both the pandemic and the war in Ukraine exposed Europe's vulnerabilities and dependency on key sectors and brought about the EU's concept of strategic autonomy, promoting self-sufficiency of the EU and giving a stronger push to the conversion to renewable energy and energy efficiency.

This race for the leadership in the green and renewable energy sectors is a welcome competition if played with fair rules further promoting the development of green technologies and leading to price reductions making these technologies and their applications affordable and accessible. Interestingly, the new government in the USA did not really "undo" the "America first" concept in terms of competition when we look at their lasting resistance to the WTO, but also in view of their catch-up strategy in the field of green technology; for the time they lost during the era of climate denial. With the adoption of the Inflation Reduction Act (IRA), the USA provides subsidies and tax cuts to attract leading green technology companies to produce in the USA. The matter turned into a major competition dispute between the USA and Europe, but it is also of concern to any other nation, in particular developing and emerging economies. Leadership in green technology has become key for competitiveness and global leadership and in international cooperation and partnerships. To promote green technology and climate change policies as well as the Sustainable Development Goals (SDGs), including the right to adequate food, are at the heart of the EU's external policies.

Latin America has the potential to play a major role in promoting a positive Climate Change agenda in particular since the re-election of Luis Ignacio Lula da Silva, who intends to win back Brazil's leadership role and re-establish regional cooperation in Latin America. At the Major Economies Forum on Energy and Climate (MEF), Lula highlighted Brazil's commitment to zero deforestation by 2030, and the partnership with other countries with tropical forests and announced a summit of all eight Amazon countries to promote a new common agenda for the region for August 2023. This could pave the way for a strengthened co-operation between Europe and Latin America particularly in times of strong global geopolitical tensions.

Environmental and democratic concerns proved to be an obstacle for the adoption of the Mercosur agreement in several national parliaments of EU Member States and in the European Parliament. It is of crucial importance to take them on board and integrating real safeguards as well as a more "inclusive" and "participatory" and "democratic" approach both in the negotiation process and certainly before the signing of agreements. From a democratic point of view, it is out of the question that any bi-regional agreement with repercussions on economic, environmental, consumer protection and social policies should come into force without the adoption of democratically elected parliaments – be it the national assemblies or the European Parliament.

Last but not least, I very much agree with Andrea Ribeiro Hoffmann and Christian Ghymers that in the current crisis of multilateralism, more intensive and strategic cooperation of Europe and Latin America on the matter of climate change might not only be the second best option but could turn out to be key to make a real change at global level and setting the basis for the future cooperation in other policy areas.

Founder of the Jean Monnet Network   Bettina de Souza Guilherme
Crisis-Equity-Democracy
for Europe and Latin America
European Parliament
Brussels, Belgium

Acknowledgements

The European Commission support for the production of this publication does not constitute an endorsement of the contents which reflects the views only of the authors, and the Commission cannot be held responsible for any use which may be made of the information contained therein.

This project has been selected as a Jean Monnet Network and funded with support from the European Commission. This publication reflects the views of the authors alone, and the Commission and the European Parliament cannot be held responsible for any use which may be made of the information contained herein.

Contents

Part III   New Green Solutions to Climate Change

Introduction

Andrea Ribeiro Hoffmann, Paula Sandrin, and Yannis E. Doukas

Climate change has been an urgent matter in international politics since at least the late 1980s when the Brundtland Commission Report, *Our Common Future*, was published (World Commission on Environment and Development, 1987) alerting to the effects of the ozone hole and defining the concept of "sustainable development" as the development that meets the needs of the present without compromising the ability of future generations to meet their own needs. The conclusion of the United Nations Framework Convention on Climate Change (UNFCCC) in 1992, resulting from the United Nations Conference on Environment and Development (UNCED) that took place in Rio de Janeiro, the "Earth Summit," was a breakthrough for international cooperation. The UNFCCC yearly Conferences of the Parties (COPs) are by now a focal point for global exchanges not only about the implementation of the agreement but also about environmental commitments broadly speaking. In the context of the COPs, two main treaties were produced, the Kyoto Protocol, in 1997 (in force in 2005), and the Paris Agreement, in 2015 (in force in 2016). A new stage of concern was achieved with the launching of the FCCC Global Warming Report of 1.5 C, in 2018, with a detailed assessment of the impact of global warming. The current global ecological crisis is so profound that it has been suggested that an epoch-scale boundary has been crossed, from the Holocene to the so-called Anthropocene; biophysical impacts include the rise in global temperature and sea levels, ocean acidification, and coral bleaching, biodiversity loss, deforestation, and water, soil and air pollution (Ramos, 2020, p. 813).

A. Ribeiro Hoffmann (✉) · P. Sandrin
Pontifical Catholic University of Rio de Janeiro, Rio de Janeiro, Brazil
e-mail: a_ribeiro_hoffmann@puc-rio.br; paula-sandrine@puc-rio.br

Y. E. Doukas
National and Kapodistrian University of Athens, Athens, Greece
e-mail: jodoukas@pspa.uoa.gr

© The Author(s) 2024   1
A. Ribeiro Hoffmann et al. (eds.), *Climate Change in Regional Perspective*,
United Nations University Series on Regionalism 27,
https://doi.org/10.1007/978-3-031-49329-4_1

Since the 1990s an ever-expanding literature has addressed the possibilities and challenges for international cooperation to address environmental problems and climate change; this book addresses these issues from a less studied perspective, namely, comparative regionalism. Comparative regionalism studies have explored the role of regions in global politics and multilateral cooperation and have initially focused on economics and security in the context of the post-Second World War (Katzenstein, 1996; Mansfield & Milner, 1997; Solingen, 2014). In the 1990s, with the deepening of regional integration in Europe, the regional level became a focus of attention on a broader range of areas, such as health, education, gender, and migration. As a research agenda, comparative regionalism studies propose to overcome the centrality of the European Union (EU) in regionalism studies and compare regional processes and organizations in the Americas, Africa, and Asia from an equal foot methodological perspective (Fawcett, 2004; Sbragia, 2008; De Lombaerde et al., 2010; Acharya, 2012; Börzel & Risse, 2016).

Comparative regionalism studies in the areas of environment and climate change are scarce. In what seems to be the first book addressing this research agenda, Elliott and Breslin (Elliot and Breslin 2011) published the edited volume *Comparative Environmental Regionalism* in 2011 including chapters on the EU environmental policy, "pan-European" cooperation, and regional initiatives in East Asia, South Asia, sub-Saharan Africa, the Middle East, and North Africa. Jorg Balsiger advanced the idea of regional environmental governance in a number of publications and special issues (Balsiger & Debarbieux, 2011; Balsiger & Vandeveer, 2012; Balsiger & Prys, 2016). The empirical focus of the 2011 and 2012 special issues was mostly Europe and Asia, especially the Association of Southeast Asian Nations (ASEAN), and the 2016 publication introduced a multidimensional typology of regional agreements to present a systematic account of regional environmental governance as a first step to further research, given what they identified then as a lack of research in this area. Schreurs (2013) contributed with a chapter on regionalism and environment governance in a *Handbook of Global Climate and Environment Policy* edited by R Falkner in which she analyzes the EU institutions for environmental protection, its leadership at the global level, and environmental cooperation in the North American Free Trade Agreement (NAFTA) and the ASEAN.

In a chapter published in the *Oxford Handbook of Comparative Regionalism*, edited by Tanja Börzel and Thomas Risse, in 2016, Peter Haas (Haas 2016) broadly defined environmental governance at the regional level as "processes of collective deliberations about norms, institutions, participation, practices and rules which occur at geographical scales associated with major conventional regions – or essentially continents or where those continents collide in an effort to address transboundary environmental degradation occurring at the regional scale" (p. 431). In the chapter, Haas analyzed several cases of regional governance of river basins, regional seas, air pollution, marine fisheries, and others, in Africa, Americas, Asia, Europe and Eurasia, and the Middle East, including governance by three formal regional organizations: NAFTA, EU, and ASEAN. In the case of the Americas, the only formal regional organization included was NAFTA, and other cases were the Comisión

Permanent del Pacífico Sur (CPPS) and the Central American Marine Transport Commission (COCATRAM).

There is a rich literature about the environment and climate change in Latin America, but no publication has addressed these areas through the lenses of comparative regionalism or compared regional European and Latin American agendas and policies on the environment and climate change, as we do in this book. The four axes that Bianculli and Ribeiro Hoffmann (2016, p.651) propose to study regional social policies are useful to think about how the environment and climate change have been addressed in Europe and Latin America and to what extent and how the regional level contributes to policymaking in the fields of environment and climate change, vis-à-vis the global and the national levels. The four axes are regional redistribution mechanisms, regional regulations, regional rights, and regional cooperation. Regional redistribution mechanisms include regional banks and funds from third parties, regional regulations can include the setting of standards to avoid a race to the bottom and the regulation of private social services, regional social rights can be assured in regional treaties, and regional cooperation includes technical cooperation, capacity building, and harmonization of domestic policies and regulations.

This edited volume has two main objectives: the first is to critically analyze and compare the role of regions in addressing environmental and climate change challenges in Europe and Latin America. The second objective is to assess the initiatives developed among these regions and formulate recommendations, contributing, therefore, to the mutual understanding of the issues at stake. The second objective is particularly important given that the book results from research developed by the Jean Monnet Network (JMN) "Crisis-Equity-Democracy" (620963-EPP-1-2020-1--BR-EPPJMO-NETWORK), which aims at strengthening the Strategic Partnership between the EU and the Community of Latin American and Caribbean States (CELAC).

Given these objectives, the book is structured in three main sections. In the first section, the authors discuss cooperation and perspectives on climate change within and among the EU and Latin American regional organizations. The first two chapters analyze the development of the norms, agendas, and initiatives discussed and implemented at the regional level, including an analysis of existing regional redistribution mechanisms, regional regulations, and regional cooperation, as well as reflections about rights-based approaches to sustainable development including the rights of nature. *Jamile Mata Diz* and *Márcio Luís de Oliveira* (chapter "The EU in a Multi-Dimensional Regime: The Regulation Of Climate Neutrality") focus on the EU and *Andrea Ribeiro Hoffmann* (chapter "Climate Change Cooperation in Latin American Regionalism") on the Southern Cone Market (MERCOSUR), the Union of South American Nations (UNASUR), the Forum for the Progress and Integration of South America (PROSUR), and CELAC. The two following chapters explore these issues at the interregional level, *Federico Castiglioni* focuses on EU-MERCOSUR relations (chapter "An "Aggressive" Cooperation: Environment as a Hot Issue in EU-LAC Relations"), and *Christian Ghymers* focuses on EU-CELAC (chapter "Fostering the Dynamics of the Bi-regional Summit EU-CELAC for Spurring the Cooperation in Climate Change").

The second section addresses the challenges to finance development and a "greener" economy in both regions, reflecting on the existing mechanisms and potential innovations, i.e., mostly the question of redistribution at the global and regional levels. *Stephany Griffith-Jones and Marco Carreras* (chapter "The Role of European Investment Bank (EIB) and National And Regional Development Banks in The Green Transformation") examine the countercyclical role of the European Investment Bank and development banks to the green transformation in the EU and at the global level, as well to the LAC region. *Stephan Schulmeister* (chapter "Fixing Rising Price Paths for Fossil Energy – Basis of a "Green Growth" Without Rebound Effects") presents an ambitious alternative approach to carbon taxes and emission trading schemes as mechanisms to incentivize the necessary investments in a permanent reduction of carbon emissions, taking the EU and its European Green Deal as an example, and consisting of fixing long-term price paths for crude oil, coal, and natural gas.

The chapters of the third section critically assess so-called new green solutions to climate change within these two regions, illustrating the challenges of fostering consensus on priorities and most appropriate mechanisms, policies, and projects on the ground. They address mostly the axes of regulation, rights, and cooperation. The first three chapters focus on the agricultural sector, and the last two take an overall perspective. *Yannis E. Doukas, Ioannis Vardopoulos, and Pavlos Petides* (chapter "Challenging the Status Quo: A Critical Analysis of the Common Agricultural Policy's Shift Towards Sustainability") explore how the Common Agricultural Policy (CAP)'s move toward "greening" is redefining the trajectory of EU and global agriculture, and *Napoleon Maravegias, Yannis E. Doukas, and Pavlos Petides* (chap. "Climate Change Concerns and the role of Research & Innovation in the Agricultural Sector: The European Union Context") focus on the more specific role of research and innovation (R&I) in the agricultural sector, to deal with climate change challenges, especially in the context of the EU and the new CAP. *Alexandra Teixeira and Camila Amorim Jardim* (chapter "Building Climate-Resilient Food Systems: The Case of IFAD in Brazil's Semiarid") turn to Latin America and analyze how climate change is worsening food insecurity and malnutrition in this region and explore regional pathways, challenges, and project-specific solutions for building sustainable climate-resilient food systems, presenting the case study of International Fund for Agricultural Development (IFAD)'s Pro-Semiarid Project in Bahia, Brazil (PSA). Last but not least, *Paula Sandrin* (chapter "EU and Brazil in the International Circuits of Disavowal of the Climate Crisis Paula Sandrin (PUC-Rio)") analyzes joint initiatives to tackle the climate crisis and the optimism surrounding green hydrogen as a possible new source of sustainable connectivity between EU and LAC in the illustrative case of EU-Brazil projects, through the prism of the psychoanalytical concept of disavowal.

In conclusion, we draw on the findings of the chapters and the four axes of social policymaking beyond nation-states, i.e., redistribution mechanisms, regulations, rights, and cooperation on environmental issues and climate change to elaborate common findings about the (potential) role of the regional level in this field and advance recommendations in view of the EU-CELAC Strategic Partnership. The case for a (re)scaling of policy decision-making and implementation from nation-states has been more accepted in the fields of environment and climate change than other social policies such as health or education, but this recognition does not imply a consensus on the diagnostic or on how to address the (common) problems. In this volume the chapters address environmental and climate change from different disciplines and theoretical perspectives, by authors from Europe and Latin America and senior and junior scholars. Most authors have been collaborating in the Jean Monnet Network "Crisis-Equity-Democracy" since 2016 when it started, and others joined later; the volume is therefore a sample of the diversity and complexity of the issues in hand. The aim of this edited volume was not to reach a consensus on all the issues we address but to provoke critical thinking in both regions, assuming that fostering a space of exchange of perspectives and dialogue among scholars and practitioners from both regions from a comparative regionalism perspective may contribute to the strengthening the EU-CELAC Strategic Partnership and the achievement of common solutions on environmental and climate change issues at a particularly critical global juncture.

References

Acharya, A. (2012). Comparative regionalism: A field whose time has come? *The International Spectator, 47*(1), 3–15.

Balsiger, J., & Debarbieux, B. (Eds.). (2011). Regional environmental governance: interdisciplinary perspectives, theoretical issues, comparative designs. *Procedia – Social and Behavioral Sciences, 14*, 1–202.

Balsiger, J., & Prys, M. (2016). Regional agreements in international environmental politics. *International Environmental Agreements: Politics, Law and Economics, 16*(2), 239–260.

Balsiger, J., & Vandeveer, S. D. (2012). Navigating regional environmental governance. Introduction to special issue. *Global Environmental Politics, 12*(3), 1–17.

Bianculli, A. C., & Hoffmann, A. R. (Eds.). (2016). Regional organizations and social policy in Europe and Latin America: a space for social citizenship?. Springer.

Börzel, T. A., & Risse, T. (2016). *The Oxford handbook of comparative regionalism.* Oxford University Press.

Haas, P. (2016). Comparative environmental regionalism. In T. A. Börzel & T. Risse (Eds.), *The Oxford handbook of comparative regionalism.* Oxford University Press.

De Lombaerde, P., Söderbaum, F., Van Langenhove, L., & Baert, F. (2010). The problem of comparison in comparative regionalism. *Review of International Studies, 36*(3), 731–753.

Elliott, L. M., & Breslin, S. (Eds.). (2011). *Comparative environmental regionalism.* Routledge.

Fawcett, L. (2004). Exploring regional domains: a comparative history of regionalism. *International Affairs, 80*(3), 429–446.

Katzenstein, P. J. (1996). Regionalism in comparative perspective. *Cooperation and Conflict, 31*(2), 123–159.

Mansfield, E. D., & Milner, H. V. (Eds.). (1997). *The political economy of regionalism.* Columbia University Press.

Ramos, G. D. (2020). International political economy and the environment. In *The Routledge handbook to global political economy* (pp. 813–827). Routledge.

Sbragia, A. (2008). Comparative regionalism: What might it be. *JCMS Journal of Common Market Studies, 46*, 29.

Schreurs, M. (2013). Regionalism and environment governance. In R. Falkner (Ed.), *The handbook of global climate and environment policy*. Wiley-Blackwell.

Solingen, E. (2014). *Comparative regionalism: Economics and security*. Routledge.

Part I
Regional Cooperation on Climate Change Within and Among the EU and Latin American Regional Organizations

The EU in a Multidimensional Regime: The Regulation of Climate Neutrality

Jamile Bergamaschine Mata Diz and Márcio Luís de Oliveira

Introduction

Climate change has long received attention from the scientific community, civil society, the private sector, governments, and international institutions, in local, regional, national, and international forums for discussions and deliberations. However, as this is a highly complex and transversal problem, climate change and its social, economic, and environmental consequences have not been the object of public policies for effective detection, elaboration, implementation, and evaluation.

In this context, several initiatives on the topic of climate change and its consequences have been adopted across the globe. The efforts of international society, although converging in several aspects, have revealed resistance from public and private interests of multiple dimensions and even difficulties in technical-scientific frameworks.

Thus, some international actors have stood out, in a more forceful way, to better understand and face the phenomenon of climate change and its effects. In this sense, the European Union (EU) has had a prominent role in the political, economic, social, and legal scenarios.

In the area of European Community law, the legal frameworks have been greatly improved to have repercussions both within the EU's internal scope and in its external relations, especially with the other states with which it maintains closer political-economic relations and with international society.

J. B. M. Diz (✉) · M. L. de Oliveira
Centre of Excellence Jean Monnet, Federal University of Minas Gerais,
Belo Horizonte, Brazil

© The Author(s) 2024
A. Ribeiro Hoffmann et al. (eds.), *Climate Change in Regional Perspective*,
United Nations University Series on Regionalism 27,
https://doi.org/10.1007/978-3-031-49329-4_2

The EU institutions and the governments of EU Member States have a full under-standing of the history of socioeconomically and environmentally unsustainable policies implemented, for centuries, in Europe itself and in all continents where the European states had a colonial intervention. However, currently, the EU has already incorporated, as an unavoidable and irrevocable commitment, the perspective of the need to change the civilizational paradigm to embrace social, economic, and envi-ronmental sustainability as a vector principle of its local, regional, national, intra-community, and international relations.

From the observation already pointed out (the need to improve the legal frame-works for confronting climate change and its consequences), this chapter is driven by the following core problem: in the context of climate change and the develop-ment of European environmental public policies, what innovations stem from the European Green Deal and the European Climate Law?

In this sense, the chapter is developed under the dogmatic decisional theoretical framework of the so-called European Green Deal (EGD). The EGD is the guiding policy of the current European Commission (presented by the President of the Commission, Ursula von der Leyen, on December 11, 2019) which is expressed in a set of initiatives, strategies, and legislative acts that, as a whole, aims to achieve a just and sustainable society and the inclusive transformation of European society and economy (Fetting, 2020).

Considering the core problem and the theoretical framework already presented, the hypothesis that is intended to be verified in this work is (i) the potential effec-tiveness and impact of the EGD and (ii) the consequent European Climate Law inside the EU and in its international relations on issues of climate change and its effects in the near future.

Specifically, the chapter aims to show how European Union regulation can affect relations with third countries, within the framework of association agreements with Latin America and especially with Mercosur.

It is noted that the Union's relationship with third countries or organizations demands a specific analysis of the regulatory impact regarding an agenda based on sustainable development, where the issue of climate change reaches a high level of priority. An example is the Action Plan (2015) resulting from the II EU-CELAC Summit that took place in June 2015,[1] which dealt with climate change, thus affect-ing EU relations with Latin America.

As a general objective, the work seeks to discuss some of the premises of the EGD as a guiding EU policy in achieving socioeconomic and environmentally sus-tainable development, especially in the context of climate change impacts. As spe-cific objectives, the article aims to (i) analyze European initiatives on climate change from the normative context with a focus on the Maastricht Treaty (1992) and (ii)

---

[1] II Summit EU-CELAC. Action Plan. Brussels, 2015, available on https://www.consilium.europa. eu/media/23757/eu-celac-action-plan.pdf, accessed on March 10, 2021.

analyze the regulatory scope and the enforcement of the European policy on climate change.

The methodologies used in the research were analytical-conceptual and dogmatic-propositional, since the study intends to analyze political and legal issues of the EGD and the consequent developments in the European Climate Law. In its primary and secondary sources, the research was based on legal scholars' writings, laws, case-law adjudication, and administrative acts and reports provided by EU institutions.

The chapter– in addition to the introduction, final considerations, and bibliographical references – was organized into three topics. The first topic addressed the legal framework on climate issues adopted by the EU from Maastricht in 2018, 2021 and 2023 regulations. The second part of the chapter was dedicated to the European Green Deal and the question of climate neutrality. The third topic focused on the European Climate Law and the enforcement of the EU climate regulatory system as well as its impact on third countries, notably on the Association Agreement with Mercosur.

European Union Initiatives on Climate Change: From Maastricht to the Treaty of Lisbon and Beyond

In the history of the EU, public policies and regulations on climate change began mainly after the Treaty of Maastricht, in the 1990s. However, from the 1980s onward, the so-called European Communities already showed concern about problems arising from the greenhouse effect.

There was a coincidence between the international and the European agendas on the issues related to climate concerns. Since the first report of the Intergovernmental Panel on Climate Change (IPCC), in 1990, and the Rio/1992 Conference (also known as the United Nations Conference for Environment and Development or the Earth Summit), that climate regulation started to gain importance with the adoption of the United Nations Framework Convention on Climate Change (UNFCCC/1994). Since then, the EU begins to establish the first actions aimed, in principle, at energy efficiency and the reduction of greenhouse effects. One of the main examples was the creation of the SAVE program[2] in 1991, whose objective was to facilitate and promote the implementation of energy efficiency policies and programs.

Previously, the statement by the Council of Ministers for Energy and Environment, on October 29, 1990, emphasized that "The European Community and Member States assume that other leading countries undertake commitments along [similar] lines and, acknowledging the targets identified by a number of Member States… are willing to take actions aiming at achieving stabilization of the total CO2 emissions

---

[2] Council Decision n. 91/565/EEC of October 29, 1991, concerning the promotion of energy efficiency in the Community (SAVE program). OJ L 307, 11.8.1991

by 2000 at 1990 level in the Community as a whole." It turns out that already at that time, the Community was already demonstrating the intention of adapting to the internationally assumed commitments, albeit in a general and ambiguous way, as pointed out by Grubb (1995, p. 43): "This falls into the pattern of 'constructively ambiguous' declarations that mark many stages of the development of climate policy, most notably the Convention itself."

From Maastricht onward, the EU took a more proactive stance in achieving the objectives set out in the UNFCC, setting goals for Member States, and recognizing the importance of adopting specific regulations so that such goals could be achieved. However, it was only in 2000, after the Kyoto Protocol, that the EU launched the European Program on Climate Change (ECCP/2000), whose main objective was to examine a wide range of sectors and instruments with the potential to reduce GHG emissions and develop common and coordinated strategies to meet Kyoto targets. It was within the framework of the ECCP that the European Emissions Trading Scheme (ETS) was introduced, with national limits for emissions from the energy and industry sectors in each Member State, as well as proposals and communications relating, for example, to energy efficiency and the use of biofuels.

However, the discussions for the implementation of the targets established in Kyoto, which was incorporated by the EU in 2004,[3] aimed at reductions for the period from 2008 to 2012. Even then a decisive consensus was not reached on the main matters that would be the object of a global regulation that could encompass measures linked to the following pillars: energy, transport, agriculture, waste, and the reduction of GHGs, among other sectors and policies.

The 20–20–20[4] package was adopted in 2007 based on three main policies: (i) reduction of greenhouse gas emissions by at least 20% in comparison to 1990 levels, (ii) a 20% share of renewable energies in final energy consumption (as well as a 10% target for renewable fuels), and (iii) 20% of savings on the projected EU final energy consumption in 2020.

In 2009, due to the failure of the COP in Copenhagen – for not specifying feasible goals to be met, in addition to the difficulties arising from the revision, at the time, of the Kyoto Protocol (McGregor, 2011) – the EU launched the "Climate and Energy" Package, which was structured into four policy reviews: (i) the revision of the 2003/87/EC Directive[5] on the community system for trading greenhouse gas emission allowances (ETS), (ii) the revision of Directive 2001/77/EC and (iii)

---

[3] Decision No 280/2004/EC of the European Parliament and of the Council of February 11, 2004, concerning a mechanism for monitoring Community greenhouse gas emissions and for implementing the Kyoto Protocol, OJ L 49, 19.02.2004, 1.

[4] European Council, Presidency Conclusions — Brussels March 8 and 9, 2007, *Council of the European Union*, 7224/1/07, 2007.

[5] Posteriorly amended by Directive 2009/29/EC of the European Parliament and of the Council of April 23, 2009, Amending Directive 2003/87/EC so as to Improve and Extend the Greenhouse Gas Emission Allowance Trading Scheme of the Community, OJ L140, 5.6.2009, P. 63.

Directive 2003/30/EC[6] concerning the use of energy from renewable sources, and (iv) the revision of the norms concerning the geological storage of carbon dioxide with the consolidation of all the instruments in a single legal document (Directive 2009/31/EC).

All these actions referred, to a greater or lesser extent, to the need to meet the targets established in Kyoto and renewed in the Paris Agreement[7] (2015) and in subsequent Conferences of the Parties (COPs). The measures also aimed to boost a level of environmental protection in the bottom to up sense, that is, to verify what would be the possible obstacles to achieving a higher level of protection from a dynamic that could intertwine with the achievement of sustainable development as an objective and value for the EU.

Thus, the EU's climate policy must be understood as a set of essential public policies to achieve not only the international commitments assumed but also to consolidate itself as a global player that leads the world's initiatives on climate change. This is the main objective established by the EU in the Treaty of Lisbon[8] and also in the Framework Action Programs[9] that establish the environmental priorities that must be the object of attention by EU agents and institutions.

Both in the institutional and normative aspects, the consideration of sustainable development must be understood as a premise whose content necessarily leads to a level of protection consistent with an "expansive" perspective that encompasses transversality as an element capable of introducing environmental sustainability in planning and implementation of public policies, even those of a common and regional nature such as those adopted by the EU. Measures aimed at mitigating, reducing, and safeguarding against the effects caused by climate emergencies[10] are political policies established by the EU in the Parliament European Resolution of 28 November, 2019, on the climate and environment emergency. The number (6) of Decision 2022/591/EU of the European Parliament and of the Council (adopted on April 6, 2022) on a General Union Environment Action Program to 2030 recognizes that environmental integration (as provided for in Article 11 of the TFEU) is a concrete norm that can be invoked to obtain the "highest level of protection" in both the vertical and horizontal dimensions, given its inclusion in all EU programs and

---

[6] Posteriorly repealed by Directive 2009/28/EC of the European Parliament and of the Council of April 23, 2009, on the promotion of the use of energy from renewable sources and amending and subsequently repealing Directives 2001/77/EC and 2003/30/EC which, per your time, it was amended by Directive 2018/2001/EU of the European Parliament and of the Council of December 11, 2018, on the promotion of the use of energy from renewable sources.

[7] For example, see number 2 of the Directive 2018/2001/EU.

[8] Article 191

[9] See, for example, Decision 2022/591/EU of the European Parliament and of the Council of 6 April 2022 on a General Union Environment Action Program to 2030

[10] European Parliament resolution of November 28, 2019, on the climate and environment emergency (2019/2930(RSP), available on https://www.europarl.europa.eu/doceo/document/TA-9-2019-0078_EN.html, accessed on February 12, 2020.

actions, as well as in the actions taken by Member States to comply with European standards.

This high level of protection also extends to the Union's foreign relations, especially when it comes to negotiating and signing association agreements. The spill over beyond the borders of the Union has a direct impact on the perspective of establishing a legal framework that must also be observed by the Latin American countries that are negotiating agreements with the Union, as is the case of Mercosur.

The Action Plan resulting from the Summit held in 2015 expressly established that the states that make up the CELAC must be linked to international efforts, within the framework of the Paris Agreement, to meet the goals established therein.

A noteworthy issue is that the subsequent action plans related to the 2016, 2017, and 2021 Summits made brief mentions on climate change issues when compared to the 2015 Plan, since the subject was treated in a collateral way and with vague content.

This dual perspective (vertical/horizontal) demands a continuous effort so that the EU can effectively comply with the vast (and often ambiguous and generic) regulatory framework on climate change. In this way both in the institutional and normative aspects, the "high level" of protection is linked not only to the provisions of a legal nature that are adopted both in the regional framework but also individually by each Member State. And it is precisely in this last point lies the difficulty in establishing a coherent and systemic normative framework that is easy to apply by all participants. The very (falsely) dichotomous relationship between economic growth and environmental protection generates uncertainties regarding the observance of goals considered ambitious adopted by the EU, especially regarding climate neutrality, as will be analyzed in the next topic.

An analysis of regulatory instruments cannot be disconnected from the arguments and discourse that European institutions have been continually emphasizing since the 1990s. An example were the targets adopted in 2007 (20–20-20 as described) which were not fully achieved, as regards GHG that only 21 Member States reached their national target in 2020.[11] In addition, despite the positive target for the use of renewable energy (estimated at 21.3%), the transport sector did not achieve the expected reduction, as "the 10% target for renewable energy in the transport sector was achieved in 2020, although only by a very small margin of 0.1 percentage point" (EEA 2021). Therefore, between the global leadership position that the EU intends to have as a spokesperson and lever of potentially higher regulatory measures and the reality that the organization itself faces, there is a long way to go.

---

[11] "Bulgaria, Cyprus, Finland, Germany, Ireland and Malta would need to use flexibilities, such as buying emission quotas from other EU countries, to comply with their legal objectives." Trends and projections in Europe, 2021. EEA Report – N. 13/21, p. 9. https://www.eea.europa.eu/publications/trends-and-projections-in-europe-2021. Accessed on September 12, 2022.

Green Deal and Climate Change: Ambitious Goals for an Uncertain Scenario

The European Green Deal – known as the Green Deal – expresses the direct result of initiatives previously consolidated within the scope of the EU which, in its historicity, has been unfolding in a trajectory based on the socioeconomic and environmentally sustainable development and integration. The enforcement of the Green Deal can be seen as an objective (Article 3 of the EU Treaty) but also as a value of the EU when sustainable development is considered a human right (Article 2 of the EU Treaty).

The discussion about a "greener Europe," therefore, is not recent and is increasingly consolidated not only in the planning, execution, and control of EU sectoral policies but also in the actions of community institutions and the Member States themselves. It is about establishing, in an increasingly deeper and continuous way, the pillars of socioeconomic and environmentally sustainable development based on concrete actions, delimited by normative frameworks capable of achieving this objective, as seen above.

In the Communication from the Commission entitled The European Green Deal,[12] there is a summary of all the aspects that were analyzed so that the Deal can really combine with the actions carried out and those in progress, seeking to create a unique space for decision-making always anchored in the premise of socioeconomic and environmental sustainability.

As the initial document that led to the creation of the Deal, the referred Communication presented important questions for the establishment of the main subjects of the Green Deal, with special mention to climate neutrality and, consequently, to climate change.

In this new political-legal context, the EU has developed several strategies and regulations that, although not directly related to greenhouse gas emission limits, influence issues related to climate change, such as energy policy and the use of renewable energies, within the framework of the 20–20-20 commitment.

Another important document adopted by the EU related to climate change is the Report called "2030 Framework for climate and energy policies".[13] The Report gave rise to the communication entitled "A strategic framework on climate and energy for the period 2020-2030," in which the Commission proposed a forward-looking strategy to be applied until 2050, in line with the "Roadmap" prepared to project until 2050, which seeks to reconcile socioeconomic growth with objectives related to coping with climate change and its consequences.

---

[12] Communication from the Commission – European Green Deal. Brussels, 11.12.2019. COM (2019) 640 final. https://eur-lex.europa.eu/legal-content/EN/TXT/HTML/?uri=CELEX:52019 DC0640&from=EN. Accessed on December 2, 2022

[13] Available at http://ec.europa.eu/clima/policies/2030/index_en.htm. Accessed on December 12, 2022

Likewise, in the Communication of 2014,[14] the Commission presents relevant data in which the progress made in this matter can be verified, highlighting the index achieved in the emission of greenhouse gases and the use of renewable energies.

Furthermore, as an integral part of the EU's path toward climate neutrality, mention should be made of the Commission Communication[15] adopted in 2018 entitled "A Clean Planet for All – EU Long-term Strategy for a prosperous, modern, competitive and sustainable economy with a neutral impact on the climate." The document points out several guidelines that must be adopted in the process of transition to a circular and sustainable economy and with the purpose of reaffirming the commitments assumed by EU in relation to the Kyoto goals and the Sustainable Development Goals (ODSs). This document shows the combination of efforts aimed at combining both the international agenda and the goals previously set by the EU, as mentioned above.

The transition to an economy guided by climate neutrality should encompass everything from the agricultural, extractive, industrial, and service sectors to the involvement and actions of civil society. The reconfiguration of socioeconomic activities will have as a parameter the performance of activities that can adapt to the new reality proposed by community institutions.

It is from the macro-conjunctural perspective that the European Green Deal was conceived and elaborated, that is, to encompass different themes focused on sustainability in its broad conception. These include aspects related to climate neutrality as a central point of the Deal. On December 11, 2019, the Commission presented "The European Green Deal" aimed at creating a roadmap to promote the transition to a circular economy based on climate neutrality, i.e., as follows:

The European Green Deal (in the following: EGD) has been developed before the economic corona pandemic to "put Europe on a pathway to a sustainable future, while leaving no one behind." The objective of the EGD is to place Europe on the trajectory of a climate neutral, circular economic system. Aspects of a fair distribution of profits and burdens play a special role, which is also made clear by the reference to an "inclusive approach" of the EGD. The focus of the EGD is thus on measures that strengthen the importance of environmental and climate protection for the innovative and economic power of the EU and its Member States on the way to climate neutrality (Hainsch et al., 2020, p. 1).

Specifically, regarding climate issues, the Deal mentions both the measures that were adopted with the consequent results, as well as proposes a change in the

---

[14] Communication by the Commission to the European Parliament, the Council, the European Economic and Social Committee and the Committee of the Regions – A strategic framework in terms of climate and energy for the period 2020–2030. Brussels, 22.1.2014 COM (2014).

[15] Communication from the Commission to the European Parliament, the European Council, the Council, the European Economic and Social Committee, the Committee of the Regions and the European Investment Bank – A Clean Planet for All The EU's long-term strategy for a prosperous, modern, competitive and climate-neutral – COM/2018/773 final, available at https://eur-lex. europa.eu/legal-content/PT/TXT/?uri=CELEX:52018DC0773. Accessed on December 22, 2022.

regulatory framework related to the climate that [...] will adopt a new EU climate change adaptation strategy, more ambitious than the current one. This is an essential measure as, despite mitigation efforts, climate change will continue to create significant pressure in Europe. It is essential to redouble efforts in terms of coping capacity, resistance, prevention, and preparation in the face of climate change. Work to adapt to climate change must continue to influence public and private investment, including in nature-based solutions. It will be important to ensure that, across the EU, investors, insurers, businesses, cities, and citizens are able to access data and create tools to integrate climate change into their risk management practices (COMISSION, 2019, p. 10).

In this sense, the proposal to revise Regulation (EU) 2018/1999 – European Climate Law – in March 2020, emerged as an ambitious plan that seeks to achieve net-zero greenhouse gas emissions in the integrated space by 2050. In addition, the proposed revision of the Regulation reinforced the parameters for achieving this goal, including, in several of its provisions, the participation of civil society, green governance, and the actions that must be taken by the Commission and other institutions so that neutrality can be achieved. This proposal was approved in the form of Regulation (EU) 2021/1119 on June 30, 2021.[16]

It was after the Green Deal, therefore, that the EU increased actions aimed at achieving climate neutrality on the same bases that had already been established in the European Climate Law, in addition to proposing a review of net-zeroing GHGs by 2050. Even though quite extensive, that regulatory framework was significantly changed, including the new goals to be discussed in the next topic of this work.

However, one cannot fail to mention the impact that the EGD generated on the European normative and institutional policies, since all actions are focused on fulfilling the objectives that were outlined in the EGD, in addition to the emergence, albeit indirectly, of new initiatives seeking to streamline the European action regarding climate change or as expressed by the European Parliament, climate emergencies.

It is expected that the multidimensional actions adopted by the EU at the local, national, regional, and international levels are in line with efforts to make climate neutrality not only a European objective but also a global one. Yet, according to the European Parliament, the EGD and its correlation with the climate issues must be "legally binding EU commitment to climate neutrality by 2050 at the latest will be a powerful tool to mobilize the necessary societal, political, economic and technological forces for the transition; strongly underlines that the transition is a shared effort of all Member States, and that every Member State must contribute to implementing climate neutrality in the EU by 2050 at the latest" (European Parliament 2021).[17]

---

[16] Regulation (EU), 2021/1119 of the European Parliament and of the Council of 30 June 2021 establishing the framework for achieving climate neutrality and amending Regulations (EC) No 401/2009 and (EU) 2018/1999 ("European Climate Law"). https://eur-lex.europa.eu/eli/reg/2021/1119/oj Accessed on December 22, 2022.

[17] European Parliament resolution of January 15, 2020, on the European Green Deal (2019/2956(RSP)). P9_ TA(2020)0005 - C 270/2. 7.7.2021. https://eur-lex.europa.eu/legal-content/EN/TXT/HTML/?uri=CELEX:52020IP0005. Accessed on December 22, 2022.

Despite the efforts made by the EU to achieve ambitious targets for the reduction of GHGs and the effective conversion of the energy matrix, the reduction of waste and the adoption of low-carbon measures have still not been possible to achieve relevant rates for combating climate emergencies. One of the reinforcements for achieving the climate neutrality goal that emerged from the EGD was precisely the revision of Regulation (EU) 2018/1999, whose review of the climate goals will be analyzed in the next topic.

The Revision of the European Climate Law and the Enforcement of the Climate Change Policy

The EU's actions on climate change built over time led to a more effective and rigid posture from the adoption of Regulation 2018/1999 of the European Parliament and of the Council on December 11, 2018. The Governance of the Union on Energy and Climate Action, also known as the European Climate Law, became a legal framework that aims precisely to meet the EU's goals for 2030 (subsequently revised in 2021) and to apply the measures set out in the 2015 Paris Agreement.

This law was a direct result of several initiatives previously established by the EU, as shown in Table 1.

It should be noted that the previous acts provided for measures related to energy policy issues, as well as regulatory aspects regarding the exploration of hydrocarbons, natural gas, diesel, etc. into climate neutrality itself. The establishment of specific commitments for the reduction of GHG emissions was given by Decision n. 406/2009/EC which provides in its Article 1 the "minimum contribution of each Member State toward the fulfillment of the Community's commitment to reduce greenhouse gas emissions over the period 2013 to 2020 in respect of greenhouse gas emissions covered by this Decision and the rules on how to make such contributions and the respective evaluation." This act, in its Annex II, determined the emission limits of greenhouse gases with a fixed percentage for each Member State in relation to the respective emissions in the year 2005. Annual appropriations, in turn, were contemplated in Commission Decision 2013/162/EU, on March 26, 2013, which established progressive limits until 2020, as seen in Table 2.

In the following years, revisions and expansions of such limits were carried out, to comply with the European Climate Strategy and the internationally assumed commitments, as described above. However, the 2018 Regulation established binding measures for Member States in terms of controlling the emission of greenhouse gases by establishing a governance mechanism and climate agreement based on five dimensions (Article 1.2): (a) Energy security, (b) Internal energy market, (c) Energy efficiency, (d) Decarbonization, and (e) Research, innovation, and competitiveness.

For each of these dimensions, specific obligations were created that refer, to a greater or lesser extent, to the quantification of the national contributions that must

Table 1 EU regulations and directives regarding climate change policies

Regulation	Directive
Regulations (EC) no. 663/2009 establishing a program to aid economic recovery by granting community financial assistance to projects in the field of energy	Directives 94/22/EC of the European Parliament and of the Council of 30 May 1994 on the conditions for granting and using authorizations for the prospection, exploration and production of hydrocarbons
Regulation (EC) n. 715/2009 of the European Parliament and of the Council of July 13, 2009, on conditions for access to the natural gas transmission network	Directive 98/70/EC of the European Parliament and of the Council of October 13, 1998, relating to the quality of petrol and diesel fuels and amending council Directive 93/12/EEC
	Directive 2009/31/EC of the European Parliament and of the Council of April 23, 2009, on the geological storage of carbon dioxide
	Directive 2009/73/EC of the European Parliament and of the Council of July 13, 2009, concerning common rules for the internal market in natural gas
	Directive 2010/31/EU of the European Parliament and of the Council of May 19, 2010, on the energy performance of buildings
	Directive 2012/27/EU of the European Parliament and of the Council of October 25, 2012, on energy efficiency, amending Directives 2009/125/EC and 2010/30/EU
	Directive 2013/30/EU of the European Parliament and of the Council of June 12, 2013, on safety of offshore oil and gas operations and amending Directive 2004/35/EC
	Directive 2009/119/EC of September 14, 2009, imposing an obligation on Member States to maintain minimum stocks of crude oil and/or petroleum products
	Directive 2018/2001/EU of the European Parliament and of the Council of December 11, 2018, on the promotion of the use of energy from renewable sources

be achieved by the year 2050 for climate neutrality and by 2030 for renewable energy sources[18] and energy efficiency in the internal market.[19]

In the revision of the European Climate Law in 2021 by the Regulation (EU) 2021/1119, the goal of climate neutrality was adopted. It should be achieved by

---

[18] See Directive 2018/2001/EU of the European Parliament and of the Council of December 11, 2018, on the promotion of the use of energy from renewable sources. https://eur-lex.europa.eu/legal-content/EN/TXT/?uri=CELEX%3A02018L2001-20220607. Accessed on December 22, 2022.

[19] Directive 2012/27/EU of the European Parliament and of the Council of October 25, 2012, on energyefficiency.https://eur-lex.europa.eu/legal-content/EN/TXT/?uri=CELEX%3A02012L0027-20210101. Accessed on December 22, 2022.

Table 2 EU regulations and directives regarding energy and greenhouse gas emissions

Directives	Decisions
Directive 2016/2284/EU of the European Parliament and of the council, of December 14, 2016, on the reduction of national emissions of certain air pollutants	Decision no 406/2009/EC of the European Parliament and of the Council of April 23, 2009, on the effort of member states to reduce their greenhouse gas emissions to meet the Community's greenhouse gas emission reduction commitments up to 2020
Directive 2003/35/EC of the European Parliament and of the Council of May 26, 2003, providing for public participation in respect of the drawing up of certain plans and programs relating to the environment and amending with regard to public participation and access to justice Council Directives 85/337/EEC and 96/61/EC	Commission Decision 2013/162/EU of March 26, 2013, which establishes the Member States' annual emission allocations for the period 2013 to 2020, in accordance with Decision no. 406/2009/ EC of the European Parliament and of the Council
Directive 2018/2001/EU of the European Parliament and of the Council of December 11, 2018, on the promotion of the use of energy from renewable sources	

2050, seeking to zero the net balance of emissions. In addition, the previous target for the internal reduction of net GHG emissions forecast for 2030 increased from 30% to 55%, considering 1990 levels, in accordance with Article 4.1 of a neutral low-carbon system, which is perfectly consistent with the EGD.

It was specifically regarding decarbonization that there was a significant change since national contributions should follow the Regulation (EU) 2018/842. The objective was to achieve a 30% reduction in the respective greenhouse gas emissions by 2030, compared to 2005 levels, for the sectors of energy, industrial processes and product use, agriculture, and waste, based on a monitoring system that had previously been established by Regulation (EU) n. 525/2013, in addition to compliance with Regulation (EU) 2018/841, which created specific rules for activities related to land use and forests.

Specifically for sectors covered by Regulation (EU) 2018/842, the target was revised in April 2023 by Regulation (EU) 2023/857.[20] The main objective is to stipulate the progressive reduction of GHG emissions until they reach 40% in 2030, compared to 2005 levels, always aiming for a global average of 55% as set out in the 2021 Regulation.

Regulation (EU) 2023/857 establishes new national limits for each Member State, with a percentage fixed in its Annex I. It should be noted that this Regulation revised the national reduction targets that had been established in 2018, with more

---

[20] Regulation (EU) 2023/857 of the European Parliament and of the Council of April 19, 2023, amending Regulation (EU) 2018/842 on binding annual greenhouse gas emission reductions by Member States from 2021 to 2030 contributing to climate action to meet commitments under the Paris Agreement, and Regulation (EU) 2018/1999. https://eur-lex.europa.eu/eli/reg/2023/857/oj. Accessed on December 22, 2022.

ambitious limits than those previously set for reducing GHG emissions. As an example, Portugal went from a percentage of −17% to −28.7%; Denmark went from −39% to −50%, as well as Germany, Sweden, Finland, and Luxembourg.

The annual appropriations must be provided by the European Commission through the adoption of an implementing act, in accordance with the linear trajectories established in Article 1.2 of Regulation (EU) 2023/857 according to a specific time frame. So, 2005 will be taken as the base year for all the effects but following the data from the national inventories from 2016 to 2018 to be applied for the calculations relating to 2021 to 2025 and, subsequently, for the calculations from 2026 to 2030. The most recent data from the national inventories will be used for the years 2021, 2022, and 2023. Therefore, the EU is concerned about establishing more effective parameters to limit what the allocations of each Member State will be, based on the information provided by the respective states according to Article 26 of the Regulation (EU) 2018/1999.

Article 26 of the 2018 Regulation was also amended by the 2023 Regulation. It establishes the obligation for Member States to submit, by January 15 of each year, the preliminary data of the GHG inventory and, by March 15 of each year, starting in 2023, the final data of the GHG inventory, following the provisions contained in the Annex V of the 2018 Regulation.

In short, there is a significant regulatory effort by the EU to achieve the goals that have been established and to obtain a commitment from Member States. However, some considerations should be made about the vast body of legislation that structures the European climate strategy:

1. Detailed regulation applied only to certain sectors: as mentioned in the preliminary provisions of the 2023 Regulation (point 11), "in certain sectors, greenhouse gas emissions have increased or remained stable." In the same vein, the Communication from the Commission on September 17, 2020, when specifying that "achieving climate neutrality requires significantly intensifying EU action in all sectors" (European Commission, 2020), in addition to highlighting that "Achieving a 55% reduction in greenhouse gas emissions will require measures in all sectors" (European Commission, 2020). The transition to a decarbonized economy will cause adjustments in all economic and productive sectors, in addition to demanding a significant financial contribution so that one can speak of a total conversion to climate neutrality. The most affected sectors according to the 2020 communication are energy, industry (notably the use of fossil fuels), transport, agriculture, and civil construction. The point-to-point regulation of each of them must always consider the targets already established by the 2021 and 2023 Regulations.

2. Difficulty in complying with the regulatory system: despite the recognition of the asymmetries in each sector and state and considering the singularities existing within the scope of the 27 Member States, the flexibility of adjustments still demands greater attention from the EU institutions to avoid noncompliance with the targets, as well as does not alter the initial climate "reduction" schedule. The

2018 Regulation already provides for flexibility instruments aimed precisely at enhancing and streamlining compliance with neutrality targets.

3. Impact of regulation on third countries: in this regard, the importance of EU measures on relations with third countries should be highlighted, notably within the scope of strategic association agreements, as is the case with Mercosur. In the Mercosur-EU Agreement, there is a chapter called "Trade and sustainable development" where the main measures to be observed by the Parties are established. The chapter has 18 articles, ranging from basic aspects such as the recognition of the main international instruments that should guide the entire decision-making process related to the Agreement to dispute settlement mechanisms that will be applied when it comes to sustainable development. In this perspective, Article 2 represents the core of the chapter by establishing rights and obligations that must be observed by the Parties regarding the level of protection and domestic regulation, embodying a normative provision of mandatory compliance. The use of terms of mandatory and non-optional observance results in hard law under this article, as those terms are used from a sense of command and not mere liberality. As an example, Article 2.2 establishes "The Party should not weaken the levels of protection afforded in domestic environmental or labor law with the intention of encouraging trade or investment." There is no doubt that this is a clause whose content is not merely dissuasive but mandatory. It is not, therefore, an "umbrella" agreement clause since, upon entering into force, it assumes the observance of all the chapters contained in the Agreement, even if there is no express mention of possible punishments derived from its noncompliance. There are no empty words that can be awarded as an inherent part of an agreement of such magnitude (Mata Diz, 2021).

Regarding climate change, as well as the necessary neutrality to mitigate the negative effects it brings, Article 6 specifically mentions the recognition by the Parties of the importance of complying with the goals and devices set by the United Nations Framework Convention on Climate Change and the Paris Agreement. Therefore, the Parties must adopt measures to promote the reduction in the emission of greenhouse gases and to mitigate the effects derived from climate change.

Besides that, according to Article 6(2)(a) TSD chapter, the Parties shall effectively implement the UNFCC and the Paris Agreement. Consistent with Article 2 PA, Parties shall promote the positive contribution of trade toward low greenhouse gas emissions and climate-resilient development and to increase the ability to adapt to the adverse impacts of climate change in a manner that does not threaten food production (Article 6 (2)(b) TSD chapter). Thus, a direct link to the agricultural sector is established (Heyl et al., 2021, p. 6).

This is a central aspect that must be considered when addressing the issue of climate neutrality and relations with third countries. If the EU intends to adopt global leadership on the issue of climate change, it must undoubtedly adopt a vanguard posture, effectively applying the goals set by it and managing to minimize the effects generated by GHG emissions in its own integrated space.

In addition, relations with other states within the framework of CELAC should also adhere to the regulatory impact adopted by the EU, whether in cooperation agreements under negotiation or even in association agreements already signed. The content of the action plans leaves no room for doubt that the theme is part of the bi-regional agenda. However, efforts to comply with international regulations have not yet been sufficient to achieve the goals established in the Paris Agreement. There is an urgent need for convergence in regulation so that consensus on such goals can be reached.

Conclusions

The quest for climate neutrality represents a permanent challenge that is difficult to solve, as it demands the adoption of suitable tools that can establish a global and general framework, as it is considered an emerging problem that affects the entire international society. In this context, states must make joint efforts to implement regulations that will minimize the effects caused by climate change, seeking to preserve existing resources for present and future generations.

The EU is a leader in proposing mechanisms that provide a fair, efficient transition that encompasses all productive sectors, especially those with the greatest impact on GHG emissions. The regulatory body of the Union has been developed over time, resulting in the adoption of normative acts that, to a greater or lesser extent, aim to implement sustainable systemic transversality in all policies, programs, and actions to meet regional goals and international commitments both by the EU itself and by the Member States.

The measures adopted by the EU range from the establishment of national limits for the reduction of GHG emissions, such as the most ambitious goal of zeroing the net balance of emissions by the year 2050, having also revised the goal adopted in Regulation 2018/1999 to reduce the general limit at the European level from 30% to 55% after the revision of the European Climate Law with the entry into force of Regulation (EU) 2021/111. National limits were also revised by Regulation (EU) 2023/857 with higher percentages that can reach 50% for some of the Member States.

The impact of the measures adopted by the EU falls not only on the Member States but also on its relations with third countries based on the objectives and values of the EU, among which, sustainable development and, consequently, the norms related to climate change. In the specific case of the agreement with Mercosur, the specific chapter dealing with the issue of sustainability mentions the issue of climate change, committing both Parties to apply a high level of environmental protection that includes climate change.

Within the framework of CELAC, the Summits mention the issues of climate changes through the action plans. However, they do not reach a definitive consensus on what would be the necessary measures for the fulfillment of international and regional commitments. In addition, EU regulation with stricter parameters than those established by the CELAC countries demands an in-depth analysis within the framework of bi-regional relations, especially when it comes to negotiating and signing bi-regional agreements, as is the case with Mercosur.

Faced with a scenario of economic and political crisis, generated not only by the pandemic but also by the effects of the conflict involving Ukraine and Russia – the main supplier of gas to Europe – the decrease in the use of fossil fuels has destabilized the efforts made so far. The strategic association developed with third countries and economic blocs, including Mercosur, provides for sustainable development in the so-called extra-commercial agenda. Thus, the Parties will be required to observe stricter norms for the conversion and greening of economic sectors. This will therefore be a crucial point in the implementation of the agreements signed by the European Union and Mercosur, as it is recognized that policies aimed at climate neutrality will be even greater in the coming decades.

References

Trends and projections in Europe 2021. EEA Report – N. 13/21, p. 9. https://www.eea.europa.eu/publications/trends-and-projections-in-europe-2021. Accessed on September 12, 2022.

Communication from the Commission – European Green Deal. Brussels, 11.12.2019. COM(2019) 640 final, https://eur-lex.europa.eu/legal-content/EN/TXT/HTML/?uri=CELEX:52019DC0640&from=EN. Accessed on December 02, 2022.

Decision No 406/2009/EC of the European Parliament and of the Council of 23 April 2009 on the effort of Member States to reduce their greenhouse gas emissions to meet the Community's greenhouse gas emission reduction commitments up to 2020. https://eur-lex.europa.eu/eli/dec/2009/406/oj. Accessed on December 22, 2022.

Directive 94/22/EC of the European Parliament and of the Council of 30 May 1994 on the conditions for granting and using authorizations for the prospection, exploration and production of hydrocarbons. https://eur-lex.europa.eu/legal-content/en/ALL/?uri=CELEX%3A31994L0022. Accessed on December 22, 2022.

Directive 98/70/EC of the European Parliament and of the Council of 13 October 1998 relating to the quality of petrol and diesel fuels and amending Council Directive 93/12/EEC. https://eur-lex.europa.eu/legal-content/en/ALL/?uri=CELEX%3A31998L0070. Accessed on December 22, 2022.

Directive 2009/31/EC of the European Parliament and of the Council of 23 April 2009 on the geological storage of carbon dioxide. https://eur-lex.europa.eu/legal-content/EN/TXT/?uri=celex%3A32009L0031. Accessed on December 22, 2022.

Directive 2009/73/EC of the European Parliament and of the Council of 13 July 2009 concerning common rules for the internal market in natural gas. https://eur-lex.europa.eu/legal-content/EN/TXT/?uri=CELEX%3A02009L0073-20220623. Accessed on December 22, 2022.

Directive 2010/31/EU of the European Parliament and of the Council of 19 May 2010 on the energy performance of buildings. https://eur-lex.europa.eu/legal-content/EN/TXT/?uri=CELEX%3A02010L0031-20210101. Accessed on December 22, 2022.

Directive 2012/27/EU of the European Parliament and of the Council of 25 October 2012 on energy efficiency, amending Directives 2009/125/EC and 2010/30/EU. https://eur-lex.europa.eu/legal-content/EN/TXT/?uri=CELEX%3A02012L0027-20210101. Accessed on December 22, 2022.

Directive 2013/30/EU of the European Parliament and of the Council of 12 June 2013 on safety of offshore oil and gas operations and amending Directive 2004/35/EC. https://eur-lex.europa.eu/legal-content/EN/TXT/?uri=CELEX%3A02013L0030-20210101. Accessed on December 22, 2022.

Directive 2018/2001/EU of the European Parliament and of the Council of 11 December 2018 on the promotion of the use of energy from renewable sources. https://eur-lex.europa.eu/legal-content/EN/TXT/?uri=CELEX%3A02018L2001-20220607. Accessed on December 22, 2022.

European Council, Presidency conclusions — Brussels 8/9. Council of the European Union, 7224/1/07, March 2007.

European Parliament resolution of 15 January 2020 on the European Green Deal (2019/2956(RSP)). P9_ TA(2020)0005 - C 270/2. 7.7.2021. https://eur-lex.europa.eu/legal-content/EN/TXT/HTM L/?uri=CELEX:52020IP0005. Accessed on December 22, 2022.

Fetting, Constanze. "The European green Deal", ESDN Report, December 2020, ESDN Office, Vienna. https://www.esdn.eu/fileadmin/ESDN_Reports/ESDN_Report_2_2020.pdf. Accessed on December 05, 2022.

Grubb, M. (1995). European climate change policy in a global context. In H. O. le Bergesen, G. Parmann, & Ø. B. Thommessen (Eds.), *Green globe yearbook of international cooperation on environment and development* (pp. 41–50). Oxford University Press.

Hainsch, Karlo et al. (2020) Make the European Green Deal real: Combining climate neutrality and economic recovery, DIW Berlin: Politikberatung kompakt, n. 153, Deutsches Institut für Wirtschaftsforschung (DIW), Berlin.

Heyl, K., Ekardt, F., Roos, P., Stubenrauch, J., & Garske, B. (2021). Free trade, environment, agriculture, and Plurilateral treaties: The ambivalent example of Mercosur, CETA, and the EU–Vietnam free trade agreement. *Sustainability, 13*(6), 3153. https://doi.org/10.3390/su13063153

Maastricht (1992) - European Union, *Treaty on European Union (Consolidated Version), Treaty of Maastricht,* 7 February 1992, Official Journal of the European Communities C 325/5; 24 December 2002.

McGregor, I. (2011). Disenfranchisement of countries and civil society at COP-15 in Copenhagen. *Global Environmental Politics, 11*(1), 1–7.

Mata Diz, J. B. (2021). Acordo Mercosul União Europeia: a sustentabilidade como foco. In: Mercosul 30 anos: passado, presente e futuro. São Leopoldo: Casa leiria v.1, p. 127–143.

Regulation (EU) 2021/1119 of the European Parliament and of the Council of 30 June 2021 establishing the framework for achieving climate neutrality and amending Regulations (EC) No 401/2009 and (EU) 2018/1999 ('European Climate Law'). https://eur-lex.europa.eu/eli/reg/2021/1119/oj. Accessed on December 22, 2022.

Regulation (EC) 715/2009 of the European Parliament and of the Council of 13 July 2009 on conditions for access to the natural gas transmission network. https://eur-lex.europa.eu/legal-content/EN/TXT/?uri=CELEX%3A02009R0715-20220701. Accessed on December 22, 2022.

Regulation (EU) 2023/857 of the European Parliament and of the Council of 19 April 2023 amending Regulation (EU) 2018/842 on binding annual greenhouse gas emission reductions by Member States from 2021 to 2030 contributing to climate action to meet commitments under the Paris Agreement, and Regulation (EU) 2018/1999. https://eur-lex.europa.eu/eli/reg/2023/857/oj. Accessed on December 22, 2022.

Climate Change Cooperation in Latin American Regionalism

Andrea Ribeiro Hoffmann

Introduction

Historically, regional organizations in Latin America have dealt with climate change only marginally. This is not a complete surprise given the role of the regional level in the discussion and implementation of social policies, but it contrasts with the centrality of this topic at the bilateral and multilateral levels of the (foreign) policy-making of Latin American countries and their participation in international regimes and agreements (Bianculli & Ribeiro Hoffmann, 2016). Despite the recent period of 'paralysis', Latin American regional organizations have been very active in the region in the last decades, and over time most of them have incorporated the language of sustainable development and have planned activities in related topics under the umbrella of environmental issues, such as the management of natural resources and environmental education; some have considered or incorporated socio-environmental conditionalities and impact assessment requirements in the allocation of funding as well as in trade agreements with third parties. However, there are very few analyses of these initiatives and their effects. Considering the perspective of a renewed cycle of regionalism ahead and at the same time the rapid deterioration of ecosystems in Latin America and the Caribbean (WMO, 2021), it is vital to deepen the understanding of the possibilities and limitations that regional organizations can play in addressing the challenges of climate change. This chapter aims to contribute to this objective by discussing the role of key regional organizations in the region, namely, the Southern Cone Market (MERCOSUR), the Union of South American Nations (UNASUR), the Forum for the Progress and Integration of South America (PROSUR), and the Community of Latin American and Caribbean States (CELAC).

A. Ribeiro Hoffmann (✉)
Pontifical Catholic University of Rio de Janeiro, Rio de Janeiro, Brazil
e-mail: a_ribeiro_hoffmann@puc-rio.br

© The Author(s) 2024
A. Ribeiro Hoffmann et al. (eds.), *Climate Change in Regional Perspective*,
United Nations University Series on Regionalism 27,
https://doi.org/10.1007/978-3-031-49329-4_3

The first section of the chapter maps and analyses the norms, agendas, and initiatives implemented by these regional organizations in the field of the environment and climate change. The second section compares the aims and achievements of these regional organizations, as well as the key actors promoting and hindering further commitments. The final section reflects on the findings and elaborates recommendations based on the premise that climate change should be a key area in a new cycle of regionalism following the period of paralysis and disintegration that culminated in the end of the decade of 2010s. The empirical research draws on the secondary literature and official documents and makes use of the concepts of path dependence and unintended consequences from historical institutionalism (Pierson, 1996; Skocpol Pierson, 2002) to understand the trajectories and promises of Latin American regional organizations to tackle environment challenges and climate change.

Latin American Regional Organizations' Environmental and Climate Change Normative Framework and Agendas

MERCOSUR

MERCOSUR was created in 1991 by the Treaty of Asunción, concluded by Argentina, Brazil, Paraguay, and Uruguay. Venezuela became a full member in 2012 but was suspended in 2016. Given its territorial conditions, one would expect that the organization would have incorporated commitments in environment from early on: Brazil alone has ca 65% of the Amazon Forest; member states have some of the biggest reserves of water, including the Aquifer Guaraní, and the rivers Amazonas, de la Plata, Paraná, and Paraguay. The thousands of kilometres of coast also make the region key to the fishery management, and the Andean Mountains are one of the richest areas in biodiversity (Vergara, 2022). The Asunción Treaty refers to the environment in its Preamble, as part of its objectives: 'Considering that the expansion of the current dimensions of their national markets, through integration, is a fundamental condition for accelerating their processes of economic development with social justice; Understanding that this objective must be achieved by making more effective use of available resources, preserving the environment, improving physical interconnections, coordinating macroeconomic policies and complementing the different sectors of the economy, based on the principles of gradualness, flexibility and balance'.[1] Albertin de Morais et al. (2012) note that this wording could have

---

[1] Asunción Treaty, free translation from the original 'Considerando que a ampliação das atuais dimensões de seus mercados nacionais, através da integração, constitui condição fundamental para acelerar seus processos de desenvolvimento econômico com justiça social; Entendendo que esse objetivo deve ser alcançado mediante o aproveitamento mais eficaz dos recursos disponíveis, a preservação do meio ambiente, o melhoramento das interconexões físicas, a coordenação de políticas macroeconômica da complementação dos diferentes setores da economia, com base no princípios de gradualidade, flexibilidade e equilíbrio'.

provided the legal basis for the creation of an environment regulatory framework to MERCOSUR, but there was no consensus for further commitments then. Stuhldreher (2012) also calls attention to underlying disagreements: 'the difficulties of coordination became evident when the Additional Protocol to the Treaty of Asunción on the Environment was not agreed upon and was not supported by Argentina' (p. 196).

These authors argue that the motivation for the gradual inclusion of environmental matters in MERCOSUR came from the engagement with the multilateral level and the realization of the UN Conference on the Environment and Development in Rio, in 1992. In this sense one of the first documents addressing the environment was the 'Canela Declaration of the Presidents of the Southern Cone Countries Prior to the United Nations Conference on Environment and Development' (*Declaração de Canela dos Presidentes dos Países do Cone-Sul Prévia à Conferência das Nações Unidas sobre Meio Ambiente e Desenvolvimento*) from 1992.

In 1994, the Common Market Group (GMC) created, through GMC Resolution 22/92, the special meeting on the environment (*Reunión Especializada del Medioambiente* (REMA)) with the objective of analysing the environmental legislation of its member states and adopting measures to environmental protection. GMC Resolution 62/93 tasked the REMA to develop a timetable for the elimination of nontariff barriers related to the environment. In 1994, REMA elaborated the Basic Directives for an Environmental Policy in MERCOSUR (*Diretrizes Básicas*), approved by the Common Market Group (CMC) as Resolution n° 10/1994. After the establishment of the permanent institutional structure of MERCOSUR by the Protocol of Ouro Preto, in 1994, MERCOSUR created the Subgroup of Environment (*Subgrupo de Trabalho em matéria ambiental* (SGT) n° 6) and extinguished REMA.

In 2001, a key legal instrument was approved, the 'MERCOSUR Framework Agreement on Environment' (*Acordo-Quadro do Meio Ambiente*) (Decision CMC n° 2, 22/06/2001), establishing the objective of achieving environmentally sustainable social economic development and stating the commitment of member states to implement the international agreements and ratify the Rio Declaration and Agenda 21. The 2001 Agreement established the objective of harmonization of national legislation, but not the creation of supranational regulation. Moreover, it defined a sectorial approach to cooperation in the following: sustainable management of natural resources (wildlife, forests, protected areas, biological diversity, biosafety, water resources, fish and aquaculture resources, soil conservation), quality of life and environmental planning (basic sanitation and potable water, urban and industrial waste, hazardous waste, dangerous substances and products, protection of the atmosphere/air quality, land use planning, urban transport, renewable and/or alternative sources of energy), environmental policy instruments (environmental legislation, economic instruments, education, information, and communication, control instruments, impact assessment, accounting, management of companies, technologies, information systems, emergencies, valuation of environmental products and services), and environmentally sustainable productive activities (ecotourism, sustainable agriculture and cattle ranching, corporate environmental management, sustainable forest management, sustainable fishing) (op.cit., 369).

In 2003, the meeting of MERCOSUR Environment Ministers (*Reunião de Ministros do Meio Ambiente do Mercosul (RMMAM)*) was created (CMC Decision

No. 19/03) upgrading the political cooperation to the ministerial level. The Ministerial Meeting and the SGT6 are until today the key institutions dealing with the environment and climate change in MERCOSUR. Currently the SGT6 has eight ad hoc groups: Ad Hoc Group on Environmental Waste Management and Post-consumer Liability, Ad Hoc Group CyMA (Competitiveness and Environment), Ad Hoc Group on Combating Desertification and Drought, Ad Hoc Group on Environmental Goods and Services, Ad Hoc Group on Biodiversity, Ad Hoc Group on Air Quality, Ad Hoc Group on Environmental Management of Chemicals and Substances, and Ad Hoc Group on the SIAM, the Integrated Information System.[2] In 2004 the Additional Protocol to the Framework Agreement on the Environment was established (CMC Decision 14/04) with the objective of guiding cooperation projects and assistance in the case of environmental emergencies (*Protocolo Adicional em Matéria de Cooperação em Emergências Ambientais*).

Despite the existence of this normative and institutional framework, Stuhldreher (2012) argues that regionalism, especially after the 2000s, had a negative effect in the environmental agenda. The lack of priority given to environmental issues such as in the 2004–2006 Work Program (CMC/Dec. N° 26/03) is a case in point. She states that it has no expression or transversal mention of the environmental commitments in economic and social policies and focused rather on cooperation programs in Science and Technology and physical and energy integration (Stuhldreher, 2012, p. 197). In fact, despite the change of approach in MERCOSUR with the 'pink tide' from a trade driven to a post-liberal perspective and the development of a social agenda in areas such as education and health, the environment and the climate change agenda were not prioritized in the organization (Briceño-Ruiz & Ribeiro Hoffmann, 2015).

Stuhldreher (2012, 196) also claims that only in the context of the preparations for the United Nations Conference on Sustainable Development Rio+20 and the Cancun Summit on Climate Change some advancements were possible: during the X Meeting of Ministers of the Environment, in 2009, guidelines were proposed for a cooperation project on adaptation to climate change, and during the XI Meeting, in 2010, it was agreed to draft a common document on the progress made since Rio 1992, as well as to encourage social movements to contribute with proposals. Monteiro et al. (2021, 4) also highlight the positive role of the preparations of global-level meetings to the establishment of commitments at the regional level, especially since 2015 with the establishment of Agenda 2030, the Sustainable Development Goals, and the Paris Agreement. Vergara (2022, 175) shows that MERCOSUR member states increased their participation in multilateral treaties and agreements in the last decade and are all now signatories of several mechanisms[3] and that despite problems of implementation, in 2017, a renewed interest

---

[2] https://ambiente.MERCOSUR.int/p_3.w_s/Grupos-Ad-Hoc-.html

[3] Such as the Convention on International Trade in Endangered Species of Wild Fauna and Flora (1951), the International Plant Protection Convention (1951), the Convention on Biological Diversity (1992), and the Vienna Convention for the Protection of the Ozone Layer (1985), including the Montreal Protocol for the Protection of the Ozone Layer, in addition to the Paris Agreement

could be perceived in the topic of climate change, with the issuing of the 'MERCOSUR Special Declaration of the Member and Associated States on the Commitment to the Paris Agreement' and the 'MERCOSUR Declaration on the Agenda 2030' (op.cit., 177). This renewed interest was, however, deeply affected by the crisis of Latin American regionalism triggered by the end of the pink tide and the COVID-19 pandemic as discussed below (Nolte & Weiffen, 2020).

In addition to the effects of its member states' participation in multilateral frameworks, the interregional negotiations with the EU have also influenced MERCOSUR's commitments and agenda on climate change, as explored by Diz & Oliveira, and Castiglioni in this volume. The concept of sustainable development and the idea of environmental conditionality in Chapter 20 of the text of the EU-MERCOSUR agreement concluded in 2019 has been controversially debated in both regions (Sanahuja, 2020; Monteiro et al., 2021; Do Amaral & Martes, 2021), but it has not been ratified until the moment of this writing.[4]

UNASUR[5]

The Constitutive Treaty of the Union of South American Nations (*Tratado Constituinte da União das Nações Sul-Americanas*) was concluded by 12 states (Argentina, Bolivia, Brasil, Chile, Colombia, Ecuador, Guiana, Paraguay, Peru, Suriname, Uruguay, and Venezuela) in Brasília, on 23 May 2008, and entered into force on 11 March 2011. Its formal institutional structure includes the Council of Heads of State and Government, the Council of Chancellors (Ministers of Foreign Affairs), the Council of Delegates, 12 Ministerial Councils, a Pro-Tempore Presidency, and a Secretary General. Climate change and the environment were not prioritized in this structure; there was no Sectorial Ministerial Council addressing

---

and the agreements referring to the marine environment such as the Convention on the Conservation of Migratory Species of Wild Animals (1983), the Basel Convention on the Control of Transboundary Movements of Hazardous Wastes and their Disposal (1989), the United Nations Convention on the Law of the Sea (1982), and the United Nations Agreement on the Conservation and Management of Straddling Fish Stocks and Highly Migratory Fish Stocks (1995).

[4] https://www1.folha.uol.com.br/ambiente/2023/02/acordo-uniao-europeia-mercosul-depende-de-compromissos-ambientais-e-texto-mais-rigido-diz-eurodeputada.shtml; https://agenciabrasil.ebc.com.br/politica/noticia/2023-01/lula-defende-mudancas-em-acordo-entre-uniao-europeia-e-mercosul

[5] Despite having been in paralysis since 2018, when 9 of its 12 member states suspended their membership or announced the intention to denounce its Treaty due to political polarization in the region and the lack of consensus to appoint a new Secretary General in 2017, Long and Suni (2022) argue that these countries could return to UNASUR without major impediments; therefore, it is worth to consider this organization. In fact, Brazilian President Lula announced the reactivation of its membership on 7 April 2023. https://oglobo.globo.com/mundo/noticia/2023/04/lula-deve-anunciar-volta-do-brasil-a-unasul-nos-100-dias-de-governo.ghtml; https://g1.globo.com/mundo/noticia/2023/04/07/o-que-e-a-unasul-e-por-que-brasil-decidiu-voltar-a-integrar-o-bloco.ghtml

these themes directly despite transversal references. Piñeros et al. (2020, 124) argue that in addition to that, the Initiative for the Integration of South American Regional Infrastructure (IIRSA) was incorporated into the South American Council for Infrastructure and Planning (COSIPLAN) in 2011, whose purpose was the development of connectivity infrastructure along the Amazon region, as well as with the use of resources and water sources for its execution, with no considerations to the environment.

Despite this negative assessment of UNASUR's activities in the area of environment and climate change, Piñeros et al. (2020) argue that Article 3 of its Constitutive Treaty has elements that support and guide the environmental discussions among the member countries, especially in l. '(g): protection of biodiversity, water resources and ecosystems as well as cooperation in disaster prevention and in combating the causes and effects of climate change and effects of climate change'. They also argue that one of the most important advances in the environmental agenda of UNASUR was the approval of the 'Guidelines for the elaboration of a regional agenda for the sovereign management of natural resources and their use for the integral development of South America', after the VI Summit of Heads of State and Government, 2012, but that it is indicative of the environmental issues considered a priority by UNASUR, namely, the mining sector, the hydrocarbon (energy) sector, and water resources (op.cit., 132), also evidenced in a publication with CEPAL from 2013. Another important benchmark was Secretary General Ernesto Samper's declaration at the Paris Conference, in 2015, defining UNASUR's strategic priorities in the area of environment and climate change, including the ratification of the Kyoto Agreement, the promotion of sustainable development through the transformation of production models, and the fulfilment of the COP21 objectives (op.cit., 132).

To summarize, UNASUR's agenda in environment and climate change was not ambitious and was clearly hindered by its contradictions. De Oliveira, Campello, and Diz (2016, 254) argue that despite positive effects of the organization activities, the absence of a common framework of environmental protection and a sound methodology to measure the impact of IIRSA/COSIPLAN infrastructure projects, for instance, led to negative effects to the environment in the region. With the process of politicization since 2016 and the shutdown in its headquarters and homepage, UNASUR became paralysed, a situation that might change with the announced return of Brazil in April 2023.

PROSUR

The establishment of PROSUR in March 2019 is directly linked to the paralysis of UNASUR. The election of centre and centre-right Presidents in Latin American countries marked the end of the 'pink tide' and a new agenda for the region, in which the isolation of Venezuela was instrumental. In this context, the Declaration of Santiago for the Renewal and Strengthening of South America (*Declaración de Santiago para la Renovación y el Fortalecimiento de América del Sur*) was concluded by Argentina (Mauricio Macri), Brazil (Jair Bolsonaro), Chile (Sebastián Piñera), Colombia (Iván Duque), Equator (Lenín Moreno), Guiana (Ambassador George Talbot), Paraguay (Mario Abdo Benítez), and Peru (Martín Vizcarra). In the meeting at Santiago del Chile were also present representatives from Bolivia (vice-chancellor Carmen Almendra), Uruguay (vice-chancellor Ariel Bergamino), and Suriname (Ambassador Edgar Armaketo in Cuba), who did not sign the declaration though; Suriname was incorporated in 2022, the same year the Chile left PROSUR.

The Declaration established PROSUR's main objectives as:

1. to strengthen and prioritize dialogue among participating countries to build a space for coordination and cooperation for greater integration and coordinated action in South America;
2. Promote the integral, inclusive, and sustainable development of participating countries to achieve greater well-being, overcome poverty, greater equality of opportunities and social inclusion, access to quality education, citizen participation and strengthening of freedoms and democracy.[6]

The institutional structure was set by the Operating Guidelines (*Liniamientos para o Funcionamento da PROSUR*), approved on the 25 September 2019, by the Ministers of External Relations of the participating countries, and includes a Presidential Summit, the Meeting of Foreign Ministers, and Sectorial National Coordination as focal points for cooperation. The Santiago Declaration established five thematic areas of cooperation in 2019: infrastructure, energy, health, defence, security and combat of crime, and disaster risk management.

Piñeros et al. (2020) argue that climate change transversalities were incorporated in the areas of infrastructure and energy, but not in other associated issues such as the automotive sector and public transportation and consumer habits of waste disposal, and that the ecological problems associated with activities such as mining, oil extraction, and large-scale agriculture are minimized or reduced to the search for best practices, without a structural discussion of the effects of extractive development models. Furthermore, he argues that in the area of disaster and risk management, emphasis is placed on promoting research, development, innovation, and entrepreneurship to increase the efficiency of disaster risk, but little is said about the risks and disasters caused by economic activities that require a fundamental transformation. The environment also does not appear in the areas of defence, citizen security, and health. As an example, the security approach prioritizes transnational organized crime, illicit trafficking of drugs, etc. but does not include crimes such as trafficking and exploitation of native species, illegal mining, illegal logging, or the

---

[6]Free translation from the original: '1. Fortalecer y priorizar el diálogo entre los países participantes para construir un espacio de coordinación y cooperación para una mayor integración y acción coordinada en América del Sur.; 2. Impulsar el desarrollo integral inclusivo y sustentable de los países participantes para lograr un mayor bienestar, la superación de la pobreza, mayor igualdad de oportunidades e inclusión social, acceso a educación de calidad, participación ciudadana y fortalecimiento de libertades y democracia' Declaración de Santiago para la Renovación y el Fortalecimiento de América del Sur.

prevention and prosecution of crimes against authorities and environmental defenders (op.cit., 144–145).

The Declaration from the 2nd Presidential Summit, which took place on 12 December, 2020, added the area of Environment and Sustainable Development to PROSUR. A Working Group Environment was created to lead work in this area, that issued a 'Sector plan for the Environment and Sustainable Development thematic area' (*Plan sectorial del área temática Medio Ambiente y Desarrollo Sostenible*)[7] as a result of the work conducted in 2021.[8] These meetings were facilitated by the Inter-American Development Bank and the Ministry of Environment and Sustainable Development of Colombia in its role of Sectorial Coordinator of the Pro-Tempore Presidency. The Plan defines three subgroups, with objectives and lines of action: sustainable consumption and production patterns (SDG 12) and promotion of the bioeconomy; sustainable transport infrastructure, and environmental education, and highlights that Brazil, Chile, Colombia, Ecuador, Guyana, Perú, and Paraguay participated more actively in the work. Not much could be found in terms of implementation of these objectives, as the region was severely reached by the COVID-19 pandemic in this period. The exit of Chile of PROSUR in 2022 and Brazil in 2023 and prospects of the revival of UNASUR have raised the question about the continuity of this organization, in a reverse process in the period when it was created.

CELAC

CELAC was established in December 2011 by the Declaration of Caracas, concluded during the simultaneous Summits of Latin America and the Caribbean on Integration and Development, and the Rio Group. CELAC includes all 33 countries from the LAC region and is defined as a mechanism of dialogue and political concertation based on consensus and the convergence of common interests to deal with common challenges. Bonilla and Álvarez (2013, 9) define CELAC as 'a deliberative space guided by the foreign policy of Latin American countries, characterized by issuing foreign policies without hegemonic pretensions, rooted in a discursive tradition that assumes logics of non-intervention, peaceful settlement of disputes, democratization of the international order and very strong images of anti-hegemonism', also as 'a mechanism for the construction of identities and strategic spaces in world politics world politics' (ibid, 8). In addition to the inclusion of Cuba and rapprochement between the subregions of South and Central America and the Caribbean, CELAC symbolizes a turning point in Mexico's foreign policy and a closer engagement of this country with regionalism in the LAC region, so far

---

[7] https://foroPROSUR.org/wp-content/uploads/2022/07/Anexo-1.-PROSUR-Plan-Sectorial-GS-Medio-Ambiente-y-Desarrollo-Sostenible.pdf

[8] https://foroPROSUR.org/wp-content/uploads/2022/07/Reporte-de-actividades-GT-Medio-Ambiente.pdf

focused on NAFTA. In addition to Mexico, Brazil and Venezuela were key drivers of the process, the latter especially, with the intention to create an alternative to US-led initiatives such as the Organization of American States (OAS) for the region to engage with the world collectively. In this sense, CELAC has established dialogues with both China and the EU early on (Ribeiro Hoffmann, 2021; Bonilla & Herrera-Vinelli, 2020).

In terms of its structure, CELAC established six main bodies, all of them taking decisions by consensus: Summit of Heads of State and Government, Meeting of Foreign Ministers, Presidency pro tempore, Meeting of National Coordinators, Specialized Meetings, and the Enlarged Troika, including the previous and subsequent presidencies pro tempore and one CARICOM member state. CELAC does not have an official website, but according to the information available at SELA's website,[9] it has no specific institutional mechanisms or agenda to deal with cooperation in the environment or climate change. These topics are, however, addressed in CELAC declaration and action plans issued by the Pro-Tempore Presidencies.[10] There is also evidence of joint statements and positions at the multilateral level such as the statement by Costa Rica at the UNFCCC COP20 in Lima in 2014 and preparations to the 2015 Paris Summit. The 2015 EU-CELAC Action Plan includes sustainable development, environment, climate change, biodiversity, and energy as significant areas of cooperation; see Castiglioni in this volume.

The most recent Action Plan from the Argentinian Pro-Tempore Presidency for 2022[11] established environmental cooperation as one of its 15 priorities. The main objective of the cooperation is to support the evaluation and follow-up of the regional reality based on the monitoring framework of the Sustainable Development Goals of the 2030 Agenda. Among the strategies set to achieve this objective are the promotion of regional dialogue platforms to foster the exchange of experiences and good environmental practices of international cooperation programs; the commission of a study on the state of the art and the main challenges facing the region in environmental matters to ECLAC, as well as a study quantifying the region's needs to finance its transition to a low-carbon and climate-resilient economy and to implement its national climate change and biodiversity policies; the establishment of regional dialogues that bring together staff responsible for international environ-

---

[9] http://s017.sela.org/CELAC/documentos/. SELA (Latin American and the Caribbean Economic System – *Sistema Economico Latinoamericano y del do Caribe*) was created in 1975 to promote economic cooperation and social development among its members, it has an administrative set in Caracas, and its Presidential Council meets once a year; the current 25 members are Argentina, Bahamas, Barbados, Belice, Bolivia, Brasil, Colombia, Cuba, Chile, Ecuador, El Salvador, Guatemala, Guyana, Haití, Honduras, México, Nicaragua, Panamá, Paraguay, Perú, República Dominicana, Suriname, Trinidad y Tobago, Uruguay, and Venezuela.

[10] The action plans are typically presented in CELACs Presidential Summits, except from 2017 to 2020 during the presidencies of El Salvador and Bolivia. Mexico and Argentina revived the Summits when assuming the rotative Presidency in 2020 and 2022, respectively, despite the crisis led by the removal of Brazil under the Presidency of Bolsonaro as discussed below.

[11] Plano de Ação da Presidência Pro-Tempore Argentina, at https://www.sela.org/es/centro-de-documentacion/base-de-datos-documental/bdd/83475/CELAC-argentina

mental negotiations and in charge of implementing environmental policy at the national level, in order to exchange experiences and best practices; and the promotion of synergies among the different regional forums such as the LAC Forum of Ministers of the Environment and the MERCOSUR Working Subgroup on the Environment, among others (Plano de Ação da Presidência Pro-Tempore Argentina 2021, item 11, pp. 6–7).

Comparing and Assessing the Role of RIOs in Tacking Climate Change

The mapping exercise of this paper provides information for a comparison of the agendas and activism of key regional organizations in Latin America in the areas of environment and climate change. From the point of view of historical institutionalism, (regional) institutions acquire certain characteristics over time that must be taken into consideration when assessing their potential effects: 'historical institutionalists take time seriously, specifying sequences and tracing transformations and processes of varying scale and temporality'. Historical institutionalists likewise analyse macro contexts and hypothesize about the combined effects of institutions and processes rather than examining just one institution or process at a time. Taken together, these three features – substantive agendas, temporal arguments, and attention to contexts and configurations – add up to a recognizable historical institutional approach (Pierson & Skocpol, 2002, 3). The concept of path dependence is particularly relevant: it 'can be a faddish term, lacking clear meaning, but in the best historical institutionalist scholarship it refers to the dynamics of self-reinforcing or positive feedback processes in a political system – what economists call "increasing returns" processes' (Skocpol & Pierson, 2002, 6). This concept is also important to conceptualize the conditions under which change (and inertia) occur; 'Historical institutionalists also employ timing and sequence arguments to focus on conjunctures – interaction effects between distinct causal sequences that become joined at particular points in time' (op.cit., 8). Institutions are developing products of struggle among unequal actors and, differently from rational approaches to institutions, lead to non-intended consequences, 'even where actors may be greatly concerned about the future in their efforts to design institutions, they operate in settings of great complexity and high uncertainty. As a consequence, they will often make mistakes' (op.cit., 14).

It was beyond the objectives of this chapter to do a systematic application of historical institutionalism to comparatively assess the aims and achievements, strengths, and weaknesses of Latin American regional organizations in the area of environment/climate change, but based on the previous mapping, summarized in Table 1 below, two main arguments are advanced: (1) Latin American regional organizations, in particular, MERCOSUR, UNASUR, PROSUR, and CELAC do not have robust legal and institutional frameworks in the areas of environment and climate change despite variance in terms of mechanisms and focuses; (2) regional organizations have had negative effects on the environment and climate change agenda due to their underlying concepts of development and related economic activities. Path dependences and unintended consequences have therefore character-ized the patterns of engagement and the attempts to include stronger environment and climate change commitments.

Regional organizations created in the first wave of regionalism such as MERCOSUR and post-pink tide such as PROSUR have a pro-free trade agenda and did include strong (socio-)environmental impact assessments and mitigation mech-anisms in their original normative. Despite the gradual inclusion of commitments in several subthemes, the environment and climate change are (still) framed as second-ary to trade liberalization and have had therefore negative effects on the climate change agenda. MERCOSUR upgraded the political profile of decision-making with the creation of a Ministerial Meeting in 2003 and has addressed the topic of climate change more directly such as in the 2017 Declaration, and PROSUR added a 6th objective and a sound Sectorial Plan in 2021, but implementation is uncertain, as the region was deeply affected by a crisis of regionalism in the late 2010s, aggra-vated by the COVID-19 pandemic. Organizations created during the pink tide and post-liberal and post-hegemonic periods, such as UNASUR, and the renewed 'Social MERCOSUR' adopted alternative approaches to development but the focus-ing on big infrastructure projects without effective mechanisms to mitigate (socio-) environmental impact have also incurred in negative unintended effects.

Finally, CELAC has included references to global-level commitments such as the SDGs, the Agenda 2030, and the Paris Agreement. While this organization does not have mandatory instruments, it can play an important role in establishing broad consensus in the region and with its external partners, particularly valuable in the current context of crisis of global-level multilateralism. As stated in its most recent Action Plan, CELAC encourages the synergies among the different regional forums such as the Forum of Ministers of the Environment of Latin America and the Caribbean, and MERCOSUR's Working Subgroup on the Environment (SGT6). The Forum of Ministers of the Environment of Latin America and the Caribbean was established in 1982 and is held every 2 years. The Forum does not take place in the context of a regional organization, but it is considered the most representative and political meeting in the region and works closely with the United Nations Environment Program (UNEP).[12]

Another relevant question is to understand who are the main actors pushing and hindering these agendas, including domestic and external state and non-state actors. It was seen that the participation of LAC countries in multilateral debates and nego-tiations such as the UN conferences (Stockholm 1972+, Rio 1992+) and the confer-ences of the parties (COPs) of the UNFCCC and the Convention on Biological Diversity (CDB) were relevant positive drivers, or favourable contexts for domestic

---

[12] https://www.unep.org/environmentassembly/forum-ministers-environment-latin-america-and-caribbean

Table 1 Main regional normative on the environment and climate change

	MERCOSUR	UNASUR	PROSUR	CELAC
Constitutional normative (treaty, declaration)	1991 – reference to the effective use of available resources and preservation of the environment in the Preamble of the Treaty of Asunción	2008 – references to the concept of sustainable development in the Preamble of the Constitutional Treaty; environment as a priority in Article 2; objective of the protection of biodiversity, prevention against climate change in Article 3 (g)	2019 – references to the concept of sustainable development in the Santiago Declaration	2011 – not addressed in the Caracas Declaration
Key institutions and normative	1992 Canela Declaration, REMA 1994 Basic Directives, SGT6 1995 Taranco Declaration 2001 Environmental Framework Agreement – objectives of achieving environmental sustainable social economic development; commitment of member states to implement international agreements; harmonization of national legislation; sectorial approach to cooperation 2003 Ministerial Meeting (RMMAM) 2010 Joint position Cancun Summit 2017 Declaration Paris Agreement and Agenda 2030	No sectorial council in the original structure 2012 guidelines (mining, energy, water) 2015 SG Samper Speech at Paris Summit	Not included in the five priorities in 2019 but added in 2020, when a WG on Environment was created 2021 sectorial plan with three focuses: sustainable consumption and production patterns (SDG 12) and promotion of the bioeconomy; sustainable transport infrastructure; and environmental education	Action Plan 2022, ref to SDGs and Agenda 2030, including conservation and sustainable use of biodiversity and the fulfilment of the ambitious goals it has set for mitigation and adaptation to climate change, including the fight against deforestation and energy transition; finance transition to a low-carbon and climate-resilient economy; synergies with MERCOSUR and the Forum of Ministers of the Environment of Latin America and the Caribbean

Source: Compiled by author

and regional dynamism, even if often with results below the expectations. Studies about global cooperation in the environment and climate change have emphasized the role of epistemic communities (Haas, 1992, 2015).

Epistemic communities are defined as 'networks of knowledge-based communities with an authoritative claim to policy-relevant knowledge within their domains of expertise. Their members share knowledge about the causation of social or physical phenomena in an area for which they have a reputation for competence as well as a common set of normative beliefs about what actions will benefit human welfare in such a domain' (Haas, 2015, 4–5). Still according to Haas, epistemic communities are often interdisciplinary, and their members must share principled and causal beliefs that provide a value-based rationale for social action and analytic reasons and explanations of behaviour, offering causal explanations for the multiple linkages among possible policy actions and desired outcomes, respectively. Moreover, they must have common notions of validity and a common policy enterprise.

Stuhldreher (2012) argues that the absence of epistemic communities has been a problem for the incorporation of a strong climate change agenda in MERCOSUR: 'Outside academia, the environment often lacked its own voice, unlike the economic and social interests expressed by the private sector, trade unions or other social organizations represented in parliaments' (p. 200). She also calls attention to the lack of clear leadership by Brazil or Argentina given the powerful role of agribusiness sectors, despite quite strong civil society: 'The case of Brazil is particularly interesting as it makes explicit its own conflicts around national sovereignty and the State's power to dispose of natural resources in order to sustain economic development, on the one hand, and global co-responsibility in environmental matters mobilized especially around the Amazon, on the other. The paradoxes faced by the Brazilian state are evident here: the more it seeks to profile itself as a power with regional leadership, the greater are the expectations of the international community, so that the country is confronted with the need to assume a pioneering role in South America and to comply with ecological and social standards' (op.cit., 201). These paradoxes were also analysed in the literature on Brazilian foreign policy and of other so-called rising states and represent a further challenge to the Latin American regionalism (Esteves et al., 2019).

Conclusions and Recommendations

The current regional historical context includes a possible 'new pink tide' (Farthing, 2023), the revival of UNASUR and CELAC, the realization of the 3rd EU-CELAC Summit in July 2023 after 8 years, and the possible conclusion of the EU-MERCOSUR agreement until the end of 2023. The broader global historical context includes uncertainties given the geopolitical competition between the US and China, and the war in Ukraine. This critical juncture could provide a space for a renewed regional approach to the environment and ambitious agenda to address climate change in Latin America. The leadership of Brazil under the new government of President

Lula might provide an additional driver to overcome path dependencies and position the climate change agenda in a top priority of regional politics and regional organizations. Initiatives such as the empowerment of the Ministry of Environment and Climate Change and the creation of a Ministry of Indigenous Peoples both led by Amazonian-born leaders, Marina Silva, and Sonia Guajajara, respectively, are cases in point.

The assessment of Latin American regional organizations' agendas and mechanisms to address climate change presented in this chapter could be expanded to include other regional organizations such as the Amazon Cooperation Treaty Organization (ACTO), a less studied organization in the literature of comparative regionalism but that has a concept of sustainable development and could play a key role in the Amazon region (Nunes, 2016; Filippi & Macedo, 2021), in addition to the Forum of Ministers of the Environment of Latin America and the Caribbean and the more traditional Andean Community (CAN), the Central American Integration System (SICA), and the Caribbean Community/CARICOM.[13] That said, and based on the analysis here developed of MERCOSUR, UNASUR, PROSUR, and CELAC, it is possible to highlight some recommendations for a renewed regional agenda for environment and climate change.

In terms of institutions and processes, it would be desirable to increase the participation of local states and non-state actors in the discussions and decision-making processes in the format of experts' councils and advisory boards. Epistemic communities could be fostered by the promotion of dialogue among experts, policymakers, and regional bureaucrats. The inclusion of socio-environmental conditionalities both at the regional and interregional levels is also seen as desirable as concepts of sovereignty and non-intervention must be softened if environment and climate change challenges are to be taken seriously. The concept of autonomy is more flexible and traditionally addressed in the literature and foreign policy approaches of Latin American countries (Miguez, 2022; Briceño-Ruiz & Simonoff, 2017; Fortin et al., 2021); it can accommodate better the claims of the global south to address historical and structural imbalances with the necessity of acknowledging interdependence and a sense of common fate at the global level. Finally, the complex current regional architecture that includes several organizations with overlapping mandates and membership should be taken into consideration and generate a division of labour. Organizations including trade liberalization such as MERCOSUR and PROSUR should harmonize their commitments. Political dialogue and consensus building (*concertación*) can take place in all levels, but CELAC should be the key aggregator of interests and positions at the global multilateral level and interregional relations with third partners such as the EU given its broader membership.

---

[13] http://www.sice.oas.org/Environment/environmentRTA_e.asp

References

Albertin de Morais, I., Albertin de Morais, F., & Mattos, R. B. B. (2012). MERCOSUR and the importance of a harmonized environmental legislation. *Brazilian Journal of International Law, 9*, 91.

Bianculli, A. C., & Ribeiro Hoffmann, A. (Eds.). (2016). *Regional organizations and social policy in Europe and Latin America: A space for social citizenship?* Springer.

Bonilla, A. S., & Álvarez, I. E. (2013). La Diplomacia de Cumbres frente al contexto internacional del nuevo multilateralismo político latinoamericano y del Caribe. In A. Bonilla Soria & I. Á. Echandi (Eds.), *Desafíos estratégicos del regionalismo contemporáneo: celac e Iberoamérica*. Flacso/aecid.

Bonilla, A. S., & Herrera-Vinelli, L. (2020). CELAC como vehículo estratégico de relacionamiento de China hacia América Latina (2011–2018). *Revista CIDOB d'Afers Internacionals, 124*, 173–198.

Briceño-Ruiz, J., & Ribeiro Hoffmann, A. (2015). Post-hegemonic regionalism, UNASUR, and the reconfiguration of regional cooperation in South America. *Canadian Journal of Latin American and Caribbean Studies, 40*(1), 48–62.

Briceño-Ruiz, J., & Simonoff, A. (2017). La Escuela de la Autonomía, América Latina y la teoría de las relaciones internacionales. *Estudios Internacionales, 2017, 49*(186), 39–89.

De Oliveira J., Campello. J., & Diz, J.B.M. (2016). A integração regional e os projetos de infraestrutura na América do Sul. *Direito e justiça*, 233–260.

Do Amaral, A., & Martes, M. M. (2021). The MERCOSUR-EU FTA and the obligation to implement the Paris agreement: An analysis from the Brazilian perspective. In *European yearbook of international economic law 2020* (pp. 387–410). Springer.

Esteves, P., Jumbert, M. G., & De Carvalho, B. (Eds.). (2019). *Status and the rise of Brazil: Global ambitions, humanitarian engagement and international challenges*. Springer Nature.

Farthing, L. (2023). Latin America's new left surge. *NACLA Report on the Americas, 55*(1), 1–4.

Filippi, E. E., & Macedo, M. V. A. (2021) A conversão do TCA em OTCA e as dificuldades remanescentes. *Tempo do mundo. n. 27 (dez. 2021)*, 191–214.

Fortin, C., Heine, J., & Ominami, C. (Eds.). (2021). *El no alineamiento activo y América Latina: Una doctrina para el nuevo siglo*. Editorial Catalonia.

Haas, P. M. (1992). Introduction: Epistemic communities and international policy coordination. *International Organization, 46*(1), 1–35.

Haas, P. M. (2015). *Epistemic communities, constructivism, and international environmental politics*. Routledge.

Long, G., & Suni, N. (2022). Hacia una nueva Unasur Vías de reactivación para una integración suramericana permanente. Center for Economic and Policy Research (CEPR), Octubre 2022.

Miguez, M. C. (2022). The concept of autonomy as an epistemic foundation? Many paths, many turns. In A. Acharya, M. Deciancio, & D. Tussie (Eds.), *Latin America in global international relations*. Routledge.

Monteiro, R. R., Profice, C. C., Grenno, F. E., & Schiavetti, A. (2021). Environmental aspects of the agreement between the European Union and MERCOSUR. *Research, Society and Development*, e489101523038, 2021 (CC BY 4.0) I ISSN 2525-3409 I https://doi.org/10.33448/rsd-v10i15.23038.

Nolte, D., & Weiffen, B. (Eds.). (2020). *Regionalism under stress: Europe and Latin America in comparative perspective*. Routledge.

Nunes, P. H. F. (2016). The Amazon Cooperation Treaty Organization: A critical analysis of the reasons behind its creation and development. *Brazilian Journal of International Law, 13*, 219.

Pierson, P. (1996). The path to European integration: A historical institutionalist analysis. *Comparative Political Studies, 29*(2), 123–163.

Piñeros, D. V., Ararat, P. P., & Amórtegui, D. G. (2020). De Unasur a Prosur: una gobernanza ambiental reducida y un legado de desaciertos para la Amazonia. In *Gobernanza multinivel de la Amazonia* (pp. 123–156). Fundación Konrad Adenauer y Escuela Superior de Administración Pública (ESAP).

Ribeiro Hoffmann, A. (2021). La CELAC: integración regional y multilateralismo global. In G. Molano-Cruz & J. Briceño-Ruiz (Eds.), *El regionalismo en América Latina después de la post-hegemonía* (Primera edición). Universidad Nacional Autónoma de México, Centro de Investigaciones sobre América Latina y el Caribe.

Sanahuja, J. A. (2020). Acordo Mercosul-UE: por uma cláusula ambiental vinculativa. *LatinoAmerica, 21*, 14 outubro.

Skocpol T, Pierson P. (2002). "Historical Institutionalism in Contemporary Political Science". In: Katznelson I, Milner HV Political Science: State of the Discipline. New York: W.W. Norton; pp. 693–721.

Stuhldreher, A. (2012). Construcción participativa del regionalismo estratégico:¿ hacia una agenda medioambiental externa del MERCOSUR? *Revista Brasileira de Política Internacional, 55,* 194–210.

Vergara, G. (2022). Multilateralismo, medioambiente y MERCOSUR. Un triángulo desafiante para la integración regional tras el acuerdo de asociación con la UNIÓN EUROPEA. *Pensiamento Proprio, 55,* 156.

World Meteorological Organization (WMO). (2021). State of the climate in Latin America and the Caribbean 2021. https://library.wmo.int/index.php?lvl=notice_display&id=22104#.ZFa4CXbMJD9

An "Aggressive" Cooperation: Environment as a Hot Issue in EU-LAC Relations

Federico Castiglioni

Introduction: A Global Ethics for a Globalized World

In the last century, the growing pace of technological breakthroughs and an increasingly stable and secure world order paved the way for large-scale investments and unprecedented human movements. In such a context, the cross-border initiatives started by individuals soared exponentially, following the traditional proclivity of businesses to invest in, or trade with, other nations. This phenomenon, known as "globalization," is today embracing nearly every aspect of life in almost every country in the world, relying on the Internet and exploiting the possibility to quickly travel and communicate. Inevitably, globalization has introduced humanity to new ethical controversies and challenges, as different cultures, economies, and political systems interact and compete with one another on a global scale. To address these ethical issues, several new approaches to ethics have been proposed.

Those who believe in the application of "universal ethics" argue that certain ethical principles and values, such as respect for human rights and dignity, should be applied across the world regardless of cultural, political, or economic differences.[1] According to this school of thought, the existence of globalization and therefore the spreading awareness that all the sentient beings will now share the same space and time would reinforce the moral norms upholding individual responsibility and even drive new research on ethics.[2] A second approach is what has been defined as "vir-

---

[1] William Rehg, *Insight and Solidarity: The Discourse Ethics of Jürgen Habermas*, University of California Press, 1997

[2] Karl-Otto Apel, *Globalization and the Need for Universal Ethics*, European Journal of Social Theory, Volume 3, Issue 2, 2016.

F. Castiglioni (✉)
Istituto Affari Internazionali, Rome, Italy
e-mail: f.castiglioni@iai.it

© The Author(s) 2024
A. Ribeiro Hoffmann et al. (eds.), *Climate Change in Regional Perspective*,
United Nations University Series on Regionalism 27,
https://doi.org/10.1007/978-3-031-49329-4_4

tue ethics", which emphasizes the cultural traits of individuals to assess their inner moral consistency rather than relying on a general moral rule applicable to all human beings. Virtue ethics is loosely connected with wider cultural relativism, a school of thought that underlines the differences between peoples and cultures and mostly denies the existence of a rational objectivity that would work as the ultimate *Grundnorm* for morality.

Cultural relativism doesn't always reject the notion of common ethics as such but confined the same to single interrelations and/or historical moments, thereby questioning its universal character.[3] There are many other possible approaches to solve the conundrum between ethics and politics, inspired by notorious political thinkers such as Hobbes (contractarianism)[4] or western philosopher like Kant and Aristotle.[5] Almost all of the competing schools of thought on modern ethics inevitably link the development (or simply definition) of a moral rule with political consequences that affect either the contemporary jus cogens or the jus gentium. This is an intended outcome, as the same existence of a "πολιτεία" – namely, the world – presupposes both internal regulations inspired by some kind of international principles consistent with that domestic order. Therefore, the same theory around the concept of "global ethics" (that often paves the way for an argument in favour of global governance) is grounded upon a theoretical premise with huge political consequences.

The supporters of *global ethics* don't deem it necessarily universal in the sense that it is always applicable beyond of time and space, even if this is the case for most of the contemporary studies on the subject (e.g. human rights theories). Nonetheless, all of the scholars who advocate for the existence of such ethics – or for its conceptual development – agree that the guiding principles of a global moral code should be considered universal, at least in the present (global) space and time. These markers rend global ethics an applied form of morality that transcends the realm of theoretical knowledge, focusing on, and translating into political action. In his famous book, *Global Ethics and Global Common Goods*, the philosopher Patrick Riordan discusses at length the concept of global ethics, closely associating the values upholding such an ethical posture (responsibility and knowledge, among others) with a set of intangible goods shared by all the sentient beings whose preservation inevitably binds all humanity together. The reflection of Riordan is original for it strives to connect the subjects of this applied form of ethics (individuals and institutions from the smallest to the largest scale) with the objects/goods over which this ethics finds application, such as health, peace, and environment.

Scholarly speculation about global ethics and the existence of intangible common goods shared by all humanity has gained traction in international politics. In fact, in 1945 the United Nations unarguably recognized the existence of such transnational goods (i.e. peace) and values (i.e. Justice) as a reason to foster international

---

[3] Neil Levy, *Moral Relativism: A Short Introduction*, Oneworld Pubns Ltd, 2002.

[4] Christopher Morris, *"A Contractarian Account of Moral Justification" In Moral Knowledge?*, New Readings in Moral Epistemology, Walter Sinnott-Armstrong and Mark Timmons, Oxford University Press, 1996.

[5] Jon Miller, *Kant and Aristotle on ethics*, Cambridge University Press, 2013.

cooperation, as expressed in the funding Charter of this organization. Although the term "ethics" does not transpire in the Charter itself, the beliefs stated at the beginning of the same are unarguably ethical in principle (e.g. aim for freedom and rights equality) and still inspire this organization.[6] More recently, the World Health Organization (WHO), the Food and Agriculture Organization (FAO), and the World Trade Organization (WTO) all explicitly recognized some global common goods in the respective areas of health, food security and international trade, using and interchanging different terms to call them. Furthermore, the UN itself just a few years ago recognized the importance of common goods in several resolutions and documents, including the 2030 Agenda for Sustainable Development, where environment was featured on the top of the list.[7]

Even though references to the concepts of global goods and ethics have recently proliferated, an institutional definition has been lacking.[8] A wide definition of a common good often employed by scholars is that of a resource (object) that is shared and accessible to all people (subjects), notwithstanding their location, ethnicity, or nationality, and that is considered essential for their well-being and survival. Some examples might be health, security, and access to outer space. Intuitively, the preservation of global environment is probably the chief public global good, both for its extrinsic nature as a precondition to enjoying the others (there cannot be health or security in a climate catastrophe) and for its intrinsic connection with human rights, such as the right to food and water.

Two Blocs, Two Different Sensitivities: A Comparative Perspective

The peculiar and sui generis institutional organization of the European Union makes it a good observatory to address the impact of environment on contemporary international relations. Like other institutions, the EU has adopted its own definition of public good as something that "give advantages to society from the provision of certain utilities and from satisfying particular wants and needs such as the eradication of disease or the elimination of pollution".[9] The EU "broadly" classified the public goods into five types: environment, health, knowledge, peace and security,

---

[6] United Nations Charter, San Francisco, 1945.

[7] The United Nations, *Transforming our world: the 2030 Agenda for Sustainable Development*, Agenda items 15 and 116, General Assembly, October 2015.

[8] In this regard, it is worth mentioning that in 2022 UNESCO introduced a distinction between generic common goods and global common goods as a guideline for future UN publications and to clear out the use of different wordings in past UN statements (https://www.iesalc.unesco.org/en/2022/04/10/public-goods-common-goods-and-global-common-goods-a-brief-explanation/ lastly consulted 14/2/2023).

[9] Mikaela Gavas, *The EU and global public goods*, Danish Institute for international studies (DIIS), Paper n. 5, Copenhagen, 2013.

and governance. The first item on this list, the emphasis on the environment, is consistent with the Union's history. When it was introduced in 1972, the environmental program of the then European Community was one of first real transnational challenges for this institution, which at the time had no power or competence over any of the other policy fields aforementioned. This competence over the environment was further expanded in title VII of the 1987 Single European Act, which stated unequivocally that "action by the Community relating to the environment shall be based on the principles that preventive action should be taken, that environmental damage should be rectified at source as a priority, and that the polluter should pay".[10]

From this principle would stem important effects, including ad hoc cooperations with third countries and other relevant organizations sharing the same "ethical" objectives. Hence, Article 130.5 of the Treaty states that "Within their respective spheres of competence, the Community and the member states shall cooperate with third countries and with the relevant international organizations. The arrangements for Community co-operation may be the subject of agreements between the Community and the third parties concerned".[11] The Single European Act created in this way a strong link between environment and foreign policy, in a time when the latter was at an early stage of development. In the following years, there was a progressive enlargement of the powers of the Community on environmental issues, which led to the establishment of a dedicated Agency in 1993.[12] Meanwhile, the institution gradually empowered with a specific capacity on this matter was the EU Commission, whose competence will be reaffirmed in the Treaty of Lisbon with reference to the EU climate policy.[13] However, the references of TFUE to foreign policy and cooperation with third parties – now comprehended in the title XX of the Treaty – were bequeathed from the Single European Act and thus remained unchanged.[14]

Mercosur became operational in those same years. Similar to the European Community but dissimilar to the newborn EU, the organization was exclusively founded on a strong intergovernmental exclusively foundation, and the only body not directly controlled by the member states was the Comisión de Comercio, whose duties were more technical than political and centred on advisory and regulatory functions.[15] Theoretically, the Comisión was the organism dealing with third parties

---

[10] European Community, Single European Act, title VII "Environment", Official Journal of the European Communities No L 169/1, Brussels, 1987.

[11] Ivi, Title VII, Article 130, paragraph 5.

[12] European Union, Regulation No 401/2009 of the European Parliament and of the Council on the European Environment Agency and the European Environment Information and Observation Network, Brussels, 2009.

[13] European Union, Treaty on the Functioning of the European Union, OJ C202/1 (TFEU), Article 114, Brussels, 2016.

[14] Ivi, Articles 191–193.

[15] Mercosur, Decisiones del Consejo del mercado común MERCOSUR, Dec. N° 08/94, 1994 (http://www.sice.oas.org/trade/mrcsrs/decisions/DEC994.asp).

in case of external negotiations but always within the limits of its mandate, which was providing assistance to the supreme Council of the member states (Consejo del Mercado Común). Mercosur was entitled to the same exclusive competence in the field of commerce and trade regulation as the European Community, for which it represented its member states wholly and exclusively.[16]

In the years that followed, the two organizations pursued divergent strategies. As the European Community evolved into the European Union, the European environmental policy was gradually improved. In this regard, the 2008 adoption of the EU climate and energy package and the commitment to the 20-20-20 targets by the Council of the European Union are significant milestones. The process resulted in the establishment of the European Commission's Directorate-General for Climate.[17] This heightened sensitivity had ramifications for EU foreign policy. During the same period, the EU sought to enhance its international stance, establishing its first-ever diplomatic service (European External Action Service (EEAS)) in 2011. The establishment of this diplomatic corps, overseen by the High Representative of the Union for Foreign Affairs and Security Policy, did not result in a centralized control of all elements of the EU's external relations. Important parts of foreign policy, like as development and trade, were instead delegated to the Commission, rather than the external service, which has traditionally held them. The EEAS was charged with high-level political representation for the Union as well as a comprehensive coordination function, which included ensuring that the EU's diplomatic ties with third countries, member states' foreign policies, and the Commission's trade and development discussions were all coordinated.[18]

Despite the seemingly Baroque settlement for the EU foreign action, divided between the Commission, the High Representative, and the member states themselves, Brussels has been trying to exercise its competences in this field in the most consistent way possible, fostering a strong bilateral and multilateral dialogue with nation-states and non-governmental entities but even more so with other twin regional organizations, such as the African Union and the Arab League. These collaborations bolstered the perception (and sometimes self-perception) of the EU as a player actively working to improve global governance.[19] The EU foreign policy had no preclusion of interlocutors (thus contemplating the possibility of union-to-state agreements) but should have prioritized the dialogue with third parties for specific goals that are highly associable with global common goods, such as the preservation

---

[16] Roberto Dominguez, Environmental governance in the EU-Latin America relationship, in "Regions and Cohesion", Bergham Journals, Volume 5: Issue 3, 2015.

[17] European Union, Regulation 1119 of the European Parliament and of the Council establishing the framework for achieving climate neutrality and amending Regulations (EC) No 401/2009 and (EU) 2018/1999 ('European Climate Law'), Brussels, 2021.

[18] For this end, the role of High Representative was intended in Lisbon as an institutional joint between the Commission and the member states sitting in the Council (double-hat).

[19] Hartmut Mayer, *The challenge of coherence and consistency in EU foreign policy*, Routledge, 2013.

of peace, economic growth, human rights, and preservation of global natural resources.[20]

Mercosur was one of the EU's natural allies in this endeavour for global shared goods, particularly the environment. Already in 1995, on the eve of the Mercosur's establishment, the European Community signed a Framework Cooperation Agreement with Mercosur on the margins of the European Council in Madrid. Within the scope of the political understanding described by this framework, the two blocs conducted three environmental forums, one in Luxembourg in 1996, one in the Netherlands in 1997, and one in Panama in 1998.[21] As a result, the EU applauded Mercosur's decision to reach a "Framework Agreement on Environment" in 2001 (which became effective in 2004). The agreement, which recalled the Rio Declaration of 1992, aimed primarily to unify the internal legislation of the signatory countries, so avoiding the establishment of carbon emission incentives/disincentives that would impede the common market.[22] In relation to international relations and fora, the 2001 agreement stated that "The States Parties shall cooperate in the fulfilment of the international environmental agreements...such cooperation may include, as appropriate, the adoption of common policies for the protection of the environment, the conservation of natural resources, the promotion of sustainable development, the issuance of joint communications on topics of common interest and the exchange of information on national positions in international environmental forums".[23] On paper, the *Framework Agreement* was indeed a remarkable step forward as it allowed the organization to issue directives on environmental policy and discuss this topic in international negotiations. Yet, in the following years, such policies and initiatives were often found inadequate.[24]

The EU-CELAC Action Plan

The EU adopted environmental provisions in its foreign relations with South America in 2003, when it signed a "Political and Cooperation Agreement" with the Andean Community, and in 2012, when the Commission negotiated an Association Agreement with Central America.[25] The same attention to environment, and specifically to deforestation and biodiversity, was devoted when the bilateral strategic

---

[20] European Union, Consolidated version of the Treaty on European Union, 13 December 2007, 2008/C 115/01), Title 5, Chapter 1 "General provisions on the Union's External Action", Brussels, 2007.

[21] European Commission, memo 98/57, 22 July 1998.

[22] Mercosur, Framework agreement on the environment of MERCOSUR, 22 June 2001.

[23] Ivi, Article 5.

[24] Alvaro Soutullo and Eduardo Gudynas, *How effective is the MERCOSUR's network of protected areas in representing South America's ecoregions?*, Cambridge University Press, 2005.

[25] Joren Selleslaghs, *EU environmental cooperation with Latin America: a critical assessment*, Conference: 11th Pan-European conference on International Relations, in Project: Diplomacy

partnership with Brazil and Mexico was established.[26] Between 2013 and 2015, the EU increased the ambitiousness of its regional cooperation by adopting the EU-CELAC Action Plan. According to this plan, to whose conception the EEAS greatly contributed, the cooperation between Brussels and the Latin America Countries (LAC) should have spanned across different policy areas, such as migration, scientific cooperation, and people-to-people exchanges.[27] The point n. 2 of the agreement mentioned sustainable development, environment, climate change, biodiversity, and energy as significant areas of cooperation.[28] In this context, the flagship cooperation program "Euroclima" has been identified as the EU's most significant investment in Latin America.[29]

The desire of both parties to include the environment as one of the key issues of discussion, however, must face a difficult reality. Today's South American environmental agenda strives to deepen the link between development and ecosystem preservation, and the initiative is driven by individual nation-state measures and their compliance with their individual international pledges, such as Mexico's INDS in 2015[30] and Chile's carbon price.[31] Mercosur and the entire Latin American region continue to struggle to balance their natural tendency to export vital raw materials with the emergence of stricter national and international environmental regulations. In fact, environmental protection has frequently led to conflict between Mercosur members and other international actors. For instance, disputes have arisen over the use and management of transboundary rivers, such as the Amazon River and the

---

Today: insights from the EU-Latin America interregional partnership, Barcelona, 13–17 September 2017.

[26] Ivi, p. 19.

[27] European Union, EU-Celac Action Plan, Brussels, 2015.

[28] More specifically: ": i) to promote the sustainable development of all countries and to support the achievement of the MDG and the other international agreements on these issues; ii) to ensure the effective implementation of the United Nations Framework Convention on Climate Change and the Kyoto Protocol, recognizing the scientific views regarding the limit for the increase in the global temperature; iii) to develop policies and instruments for adaptation and mitigation, to address the adverse effects of climate change and enhance long-term cooperation initiatives and to reduce the vulnerability to natural disasters; iv) to support activities oriented to reduce intensity of greenhouse gas emissions in consumption and production activities in our countries, according to existing international commitments; v) to facilitate access to and exchange of information related to best environmental practices and technologies; vi) to ensure and support the full implementation of the three objectives of the Convention on Biological Diversity; vii) to improve energy efficiency and saving as well as accessibility; viii) to develop and to deploy renewable energies and to promote energy interconnection networks, ensuring the diversification and complementarity of the energy matrix"; ibidem.

[29] Allegedly, in the following 5 years, this fund brought in Latin America 300 millions of investments to sustain energy transition and climate resilience (Joren Selleslagh, *EU environmental cooperation with Latin America: a critical assessment*, ivi, p. 28).

[30] Government of Mexico, Intended Nationally Determined Contribution, UNFCC, 2015.

[31] Rocío Román, José M. Cansino, Manuel Ordóñez, *An assessment of the effects of the new carbon tax in Chile*, in Environmental and Economic Impacts of Decarbonization, Routledge, 2017, pp. 291–311.

Paraná River, which flow across multiple national borders and are sources of pure water and hydroelectric power.[32] Despite the good intentions enshrined in the Amazon Cooperation Treaty Organization (OTCA) of 1978, deforestation of the Amazon Forest and the conversion of natural habitats into agricultural lands and pastures are a further source of diplomatic disputes and international criticism.[33]

These issues emerged in the case of the EU-CELAC plan. One of the expected outcomes of action plan was the signature of a deal to boost the commercial exchanges between the two regions, and Mercosur was recognized as the most likely partner in these negotiations. The goods that Latin America would export included agricultural commodities such as soybeans, corn, and wheat, which are in high demand in Europe. But even more, the import would encompass natural resources and raw materials that the EU common market is angry for, such as copper, timber, iron, and lithium, a key component for renewable batteries and other high-tech products. The imports from Europe would be rather focused on chemicals, luxury items, medicines, and machinery. At first, the parties considered such a deal to be mutually beneficial. However, as the content of the agreement was being discussed, it became increasingly controversial for the disastrous environmental consequences it may have caused in Latin America. The root cause of this controversy lied in a moral hazard concerning the kind of commodities that Mercosur should have exported, which was seemingly a contradiction, since both organizations somehow recognized the environment as a "global common good" to be enjoyed by their member states and the international community alike.

Still, it was undeniable that the trade of these same commodities over the last years has been causing significative disruptions of the regional natural resources in Latin America (which accounts for 3% of the world's forests, 31% of the world's freshwater, and approximately 70% of the world's species), and their further exploitation was likely to worsen the trend.[34] Agriculture and herding and in particular the production of beef and wheat are among the major causes of deforestation in Latin America, as they are frequently linked to wildfires set by criminals or greedy entrepreneurs, sometimes with the tacit approval of institutions.[35] Even more disruptive for the forests are the activities associated with timbering. In 2018, Latin America accounted for more than 10% of the world production of timber, and, although the EU has already been working substantially on sustainable lodging and forest management, Greenpeace contends that from the two biggest timber-producing regions of Brazil (Pará and Mato Grosso), more than half of the wood may come from illegal logging. As far as the extraction of raw materials is concerned, this is one of the most polluting activities known in the world today, even considering the unavoid-

---

[32] David Hill, Peru's mega-dam projects threaten Amazon River source and ecosystem collapse, Mongabay, 28 April 2015 (https://news.mongabay.com/2015/04/perus-mega-dam-projects-threaten-amazon-river-source-and-ecosystem-collapse/).

[33] Treaty for Amazonian Cooperation (OCTA), Brasilia, 1995.

[34] Ivi, p. 5.

[35] Beatriz Garcia and Laurent Pauwels, *The Promise of Cooperation in Latin America: Building Deforestation-Free Supply Chains*, Cambridge University Press, 2022.

able processing needed to rinse the metals and closely entangled with the extraction itself.[36]

The situation appears particularly concerning for the industry of lithium, a valuable resource that requires an enormous amount of water to be properly plucked out from the soil. The presence of lithium in several arid lands in South America has already doomed local communities and produced huge contamination of flora and fauna.[37] Therefore, the conundrum that both blocs should have resolved before deepening negotiations was the trade-off between a profitable exchange of tangible goods (raw materials and agricultural commodities) and a nonprofitable exchange of intangible goods (fresh air and water). The EU was expected to take the major responsibility for this gap, being the actor more committed to environmental protection both domestically and internationally. In contrast with Mercosur, the Union built its own foreign policy identity on an ethic ground, portraying itself as a defender of global common goods in general and environment in particular. As a result, only in the last two decades, the Union was engaged in a "climate diplomacy" that resulted in significant deals struck with China (Bilateral Climate Change Agreement), Japan (Environmental Partnership), Canada (CETA environmental provisions), and other ambitious commitments such as the COP26.[38] Such impressive record was expected to translate in similar provisions for Latin America.

The EU-Mercosur Trade Agreement

Shortly before the signature of the draft agreement between the EU and Mercosur, the auspices were not good. In August 2019, the heads of Ireland and France threatened to veto the agreement if Brazil did not fulfil its environmental commitments. The same year, in a letter to the then EU High Representative Federica Mogherini, Finland suggested to ban imports of Brazilian beef and exclude this commodity from the contents' deal.[39] The opposition of several environmental organizations in Europe was joined – in this case – by European associations of farmers who were rather afraid of possible repercussions over the price of meat. However, in the end, the European Union and Mercosur signed a first version on the terms initially foreseen, after two decades of long negotiations. The deal created one of the world's

---

[36] Andreas Manhart et al., *The environmental criticality of primary raw materials – A new methodology to assess global environmental hazard potentials of minerals and metals from mining*, Springer, 2018.

[37] Elena Giglio, *Extractivism and its socio-environmental impact in South America. Overview of the "lithium triangle"*, America Critica, Vol. 5 N. 1, 2021.

[38] John Vogler, *The European Contribution to Global Environmental Governance*, International Affairs (Royal Institute of International Affairs 1944-), Vol. 81, No. 4, Britain and Europe: Continuity and Change, July 2005, pp. 835–850.

[39] Francesca Colli, *The EU-Mercosur agreement: towards integrated climate policy?*, European policy brief, n. 57, Egmont Institute, 2019.

largest free trade areas, covering over 780 million people and almost a third of the global economy. Some key points of the agreement included (1) tariff reductions/eliminations on a wide range of goods, including agricultural products, automobiles, and textiles; (2) improved access to each other's service markets; (3) increased protection of geographical indications, intellectual property rights, and sustainable development; and (4) abolition of technical barriers to trade – e.g. sanitary.

The agreement outlined obligations and commitments of both parties in the areas of environmental protection, biodiversity conservation, and sustainable use of natural resources. The agreement also established a Joint Committee on Environment and Sustainable Development to monitor and review the implementation of these environmental provisions. In principle, the Agreement contains an explicit reference to a range of sustainable development goals to be achieved by both the EU and Mercosur countries.[40] As in the free agreements with Mexico and Japan, there was compelling attention to respect the present international environmental agreements and promote corporate social responsibility. Furthermore, the parties committed to not lower their current health or labour standards in order to attract trade and investments. In case of a breach, a dispute settlement procedure may be invoked by either party.[41] The agreement addressed directly some illicit activities, such as illegal trade in wildlife, (7.2c), illegal logging (8.2c), abusive fisheries, and protection of the marine environment (9.2). Yet, as its opposers underlined, the Agreement lacked of the necessary measures to legally enforce such provisions, excluding those generically ensuring the fulfilment of obligations in international Treaties.[42]

As of 2024, the EU-Mercosur agreement has not been ratified by the European Parliament, and after years even the support of the member states is wavering.[43] Notwithstanding, in October 2022, the High Representative Borrell reaffirmed the commitment of the EU institutions to harbour the agreement and push for a fast ratification, going as far as to define 2023 the "Year of Europe in Latin America, and of Latin America in Europe".[44] This statement preceded the January diplomatic trip of the German Chancellor Scholz in different Latin American countries.[45] From a

---

[40] European Union (DG Trade), The EU-Mercosur Trade Agreement, paragraph 14 "Trade and Sustainable Development" (https://policy.trade.ec.europa.eu/eu-trade-relationships-country-and-region/countries-and-regions/mercosur/eu-mercosur-agreement/text-agreement_en – lastly consulted 14 February 2023).

[41] Luca Pantaleo and Francesco Seatzu, *The Eu-Mercosur Trade Agreement: the Beginning of a New Era for the Eu-Mercosur Relations?*, Il diritto dell'Unione Europea, Fascicolo 2, 2021, pp. 315–350.

[42] Guillaume Van der Loo, *"Mixed" feelings about the EU–Mercosur deal: How to leverage it for sustainable development*, European Policy Centre – commentary, 14 April 2021.

[43] Detlef Nolte, *A last chance at an EU-Mercosur agreement*, IPSO, 6/2/2023 (https://www.ips-journal.eu/topics/economy-and-ecology/a-last-chance-at-an-eu-mercosur-agreement-6489/ lastly consulted 14 February 2023).

[44] European External Action Service (EEAS), *"Road 2023": paving the way towards a European Union-Latin America stronger partnership*, EU Strategic Communications, 31 October 2022.

[45] DW Press, *Germany's Olaf Scholz kicks off South America trip*, 28 January 2023 (https://www.dw.com/en/germanys-olaf-scholz-kicks-off-south-america-trip/a-64545066 lastly consulted 14 February 2023).

strategic viewpoint, the European stance is perfectly understandable. Many European companies are engaged in infrastructure projects in many countries across the region and notably Uruguay. Furthermore, the European investments are growing everywhere, particularly in Argentina, Bolivia, and Chile –three countries forming the so-called lithium triangle. These countries are increasingly important for the EU industry in a time marked by the post-COVID disruption of the Asian supply chain. Without the South American supply, the EU member states, already weakened by the energy crisis started after the 2022 invasion of Ukraine, could never afford the energy transition lingering at the peak of the European political agenda nor invest in high technology.

Besides, with the victory of President Lula da Silva in Brazil, the transatlantic political dialogue improved, despite some occasional frictions that cast a shadow over the reliability of the newly elected Brazilian government on environmental issues.[46] The future of the EU-Mercosur agreement lies in the ability of the EU Commission to persuade the French government to change opinion about the feasibility of this deal. However, the diplomatic weapons at EU's disposal to achieve this goal are limited, and even additional protocols to the existing agreements could induce the perception that the Union is undertaking an "environmental aggressive" course of action.[47] Considering the past experience relating to agreements with Latin America (Central America or the Andean Community), which have still not been fully ratified on the European side, the biggest challenge will be to find a compromise without upending the main points of the provisional text. Mercosur, for its part, is far from dormant, and some of its members appear ready to raise the stakes by proposing to China the same agreement that is being discussed with the EU.[48]

The EU-Mercosur agreement is still pending. Commission Von der Leyen tried to resurrect the agreement and overcome the opposition of some member states and the concern of numerous civil society organizations by proposing a "EU-Mercosur sustainability instrument" to increase the environmental accountability of Latin American nations. However, the provisional text of the Commission, which presumably leaked in March 2023, listed a number of initiatives that anti-Mercosur activists deemed insufficient. Several EU member states and civil society organizations shared the same scepticism regarding the new instrument proposed by the Commission. Austria was the first EU member state to publish a critical annex highlighting the shortcomings of the proposal, particularly in regard to food safety and agricultural imports, before the Council of the European Union for agriculture

---

[46] Euractiv, *Brazil sinks rusting old aircraft carrier in the Atlantic*, 6 February 2023 (https://www.euractiv.com/section/energy-environment/news/brazil-sinks-rusting-old-aircraft-carrier-in-the-atlantic/).

[47] In the last years, there has been a scholarly debate around the idea of "climate imperialism" or "climate colonialism" declined as an hidden cost dropped from the wealthiest countries to the Global South in order to support energy transition.

[48] Reuters, *Brazil's Lula proposes Mercosur trade deal with China after EU accord*, 25 January 2023 (https://www.reuters.com/world/americas/brazils-lula-eyes-trade-deal-between-mercosur-china-2023-01-25/ lastly consulted 14 February 2023).

convened on March 20 to discuss the impact of the agreement.[49] Yet, the Austrian delegation was not alone in raising the stakes at the meeting, as the hardliners were joined by more than ten EU ministers.[50] On the other side, it appears that today even the Mercosur countries might be reluctant to reopen the nearly finalized agreement and could reject this additional burden.[51]

Conclusions

Climate is a transnational issue and currently one of the most important topics in international relations in the global age. Given its nature, which is not limited to national boundaries, it seems appropriate to approach it from the empirical perspective of regional organizations, taking into account the attention that every international and supranational institution is devoting to the subject. In this regard, it is telling that contemporary interregional politics (particularly when it pertains to the EU external relations) focuses on environmental sustainability as inextricably linked to development and trade, thereby utilizing a holistic approach. Regionally, this entanglement is plainly evident in the EU-CELAC Action Plan, wherein commerce, international issues, and ethical concerns are inextricably interwoven. In a comparative analysis between Europe and Latin America, this chapter compared the most solid (as of today) and mutually recognizing organizations operating in the two regions, namely, the EU and Mercosur. The two actors translate their pro-environment policy in a different manner, with similar aspirations but also many differences in the practical implementation of the respective agendas, owing in part to the two organizations' distinct nature and legal bases. The EU addressed the issue of decarbonization and climate change in different fora and introduced specific provisions related to climate in its recent negotiations with Canada, Japan, and other regional blocs. By contrast, Mercosur appears to be less inclined, at least for the time being, to present climate protection as a precondition to advancing trade agreements. When the first EU-Mercosur trade deal was proposed in the early 2000s, the reaction of the European civil society community was obstructive. The preoccupation about a potential environmental hazard caused by increased pollution as a consequence of the deal slowed the negotiations until 2019, when the first round of negotiations finally ended and a draft was made public. Yet as of 2023, the provisional text is still waiting for approval from the European Parliament and the EU

---

[49] Council of the European Union, *Negotiation of the EU-Mercosur Association Agreement and agricultural Implications and information from the Austrian delegation*, Brussels, 15 March 2023.

[50] Julia Dahm, *EU agri ministers push back against Mercosur deal*, Euractiv, 10 March 2023 (https://www.euractiv.com/section/agriculture-food/news/eu-agri-ministers-push-back-against-mercosur-deal/).

[51] Andy Bounds, *EU trade deal with South America delayed by row over environmental rules*, Financial Times, 5 May 2023.

member states, who are voicing concerns on behalf of European businesses, citizens, and NGOs.

The role that environmental issues play in the EU-Mercosur relationship can be described as contradictory. On the one hand, the EU is trying to reconcile commercial and political interests with its aim to act as a normative power in global governance. The challenge is not limited to the EU-Mercosur agreement but also encompasses other interregional strategic partnerships that for expediency could not find room in the present research.[52] From a different standpoint, Mercosur and all the regions of the so-called Global South could see Europe's efforts to ensure a binding enforcement of the clauses related to the environment as a foray into their domestic affairs and a not-acceptable attempt to bring in cheap materials and impose tighter environmental standards. The EU's apparent moral superiority, which appears to distrust Latin America (though with some justification), and the insistence on mandatory environmental provisions to be included in the Mercosur agreement suggested to some scholars that the EU may try to act as a "climate imperialist" power with its partners.[53]

There are different possible solutions to this conundrum. One that has been proposed is to monetize the trade of intangible goods and connect their value to the actual trade of tangible goods. The consequence would be a new market under the control of a third regulatory party that would interfere with the prices of the traded goods by swapping them with the pollution that their production has caused. Unfortunately, the difficulty in operationalizing such a market stems from a lack of information about the actual supply chain of the goods in question, which is frequently hidden or difficult to obtain from the source. Another possibility is to invest in a bottom-up approach, improving the social responsibility of business actors by rewarding those who can demonstrate the smallest carbon footprint. Allegedly, the implementation of these measures could not only unlock the EU-Mercosur deal but also improve the international standing of the two organizations, making them compliant with the moral responsibilities that global governance entails. But the fundamental point that needs to be addressed in this instance is the responsibility of these actors to preserve the so-called global common goods and thus focus on the "objects" of immaterial trade from the ethical assumption that these organizations should exercise a greater leadership than their same membership may suggest.[54] However, this assumption can be effective only as long as it is shared across the actors of the international relations, who should ultimately accept the consequences stemming from this moral premise.

---

[52] Simon Lightfoot, *Environment and climate change in the context of EU-Africa relations*, in The Routledge Handbook of EU-Africa Relations, Chapter n. 22, Routledge, 2020.

[53] Stavros Afionis and Lindsay C. Stringer, *The environment as a strategic priority in the European Union–Brazil partnership: is the EU behaving as a normative power or soft imperialist?*, International Environmental Agreements: Politics, Law and Economics, volume 14, 2014.

[54] Charlotte Epstein, common but differentiated responsibilities, Britannica (https://www.britannica.com/topic/common-but-differentiated-responsibilities).

References

Afionis, S., & Stringer, L. C. (2014). The environment as a strategic priority in the European Union–Brazil partnership: is the EU behaving as a normative power or soft imperialist? *International Environmental Agreements: Politics, Law and Economics, 14*, 47.

Colli, F. (2019). *The EU-Mercosur agreement: Towards integrated climate policy?* (European policy brief, n. 57). Egmont Institute.

Council of the European Union. (2023, March 15). *Negotiation of the EU-Mercosur Association agreement and agricultural implications and information from the Austrian delegation*, Agri 137, Brussels.

Dominguez, R. (2015). Environmental governance in the EU-Latin America relationship, in "Regions and Cohesion". *Bergham Journals, 5*(3), 63.

European Community. (1987). Single European Act, Brussels.

European Union. (2009). Regulation no 401/2009 of the European Parliament and of the Council on the European Environment Agency and the European Environment Information and Observation Network, Brussels.

European Union. (2015). EU-Celac action plan, Brussels.

European Union. (2021). Regulation 1119 of the European Parliament and of the council establishing the framework for achieving climate neutrality and amending regulations (EC) No 401/2009 and (EU) 2018/1999 ('European Climate Law'), Brussels.

Garcia, B., & Pauwels, L. (2022). *The promise of cooperation in Latin America: Building deforestation-free supply chains*. Cambridge University Press.

Gavas, M. (2013). *The EU and global public goods*. Danish Institute for International Studies (DIIS), Paper n. 5, Copenhagen.

Giglio, E. (2021). Extractivism and its socio-environmental impact in South America. Overview of the "lithium triangle". *America Critica, 5*(1), 47–53.

Lightfoot, S. (2020). Environment and climate change in the context of EU-Africa relations. In *The Routledge handbook of EU-Africa relations*. Routledge.

Manhart, A., et al. (2018). *The environmental criticality of primary raw materials – A new methodology to assess global environmental hazard potentials of minerals and metals from mining*. Springer.

Mayer, H. (2013). *The challenge of coherence and consistency in EU foreign policy*. Routledge.

Mercosur. (2001, June 22). Framework agreement on the environment of MERCOSUR.

Pantaleo, L., & Seatzu, F. (2021). The Eu-Mercosur trade agreement: The beginning of a new era for the Eu-Mercosur relations? *Il diritto dell'Unione Europea, Fascicolo, 2.*

Román, R., Cansino, J. M., & Ordóñez, M. (2017). An assessment of the effects of the new carbon tax in Chile. In *Environmental and economic impacts of decarbonization*. Routledge.

Selleslaghs, J. (2017). EU environmental cooperation with Latin America: A critical assessment. In *Conference: 11th Pan-European conference on international relations, in project: Diplomacy today: Insights from the EU-Latin America interregional partnership*, Barcelona, 13–17 September 2017.

Soutullo, A., & Gudynas, E. (2005). *How effective is the MERCOSUR's network of protected areas in representing South America's ecoregions?* Cambridge University Press.

The United Nations. (2015, October). *Transforming our world: The 2030 agenda for sustainable development*, Agenda items 15 and 116, General Assembly

Treaty for Amazonian Cooperation (OCTA). (1995). Brasilia.

Van der Loo, G. (2021, April 14). *'Mixed' feelings about the EU–Mercosur deal: How to leverage it for sustainable development*. European Policy Centre – Commentary.

Vogler, J. (2005, July). The European contribution to global environmental governance. *International Affairs (Royal Institute of International Affairs 1944-), 81*(4), Britain and Europe: Continuity and Change.

Fostering the Dynamics of the Bi-regional Summit EU-CELAC for Spurring the Cooperation in Climate Change

Christian Ghymers

Identifying the Global Dimension of the Systemic Nature of the Climate Change

Climate change is by nature an emblematic issue of the kind of challenge that could only be met at multilateral level, not only for being a global threat of national spillovers without multilateral power but for its deeper systemic character being exposed to the "tragedy of the commons[1]". Although it is not the only problem to present simultaneously a global and a systemic feature, it is probably – except the case of a nuclear war – the most irreversible one since "we only have one Earth" (Dubos & Ward, 1972), giving to the CO_2 the same status of "weapons of mass destruction" as nuclear bombs. The risks of catastrophic effects of global warming are the object of a resilient systemic denial that must be changed very urgently for preventing a catastrophic collapse of civilizations.

Global warming is essentially the result of human activity, and the phenomena – named only in 2000 the Anthropocene (Crutzen & Stoermer, 2000)[2] for getting

---

[1] A combination of a "moral hazard" with a "prisoner's dilemma" situation in which a common good is a finite resource and is freely available no one has an incentive to preserve or to reinvest in maintaining the good since each agent acts in his own self-interest because he cannot prevent others from appropriating the value of the investment by consuming the product for themselves. The good becomes more and more scarce and may end up entirely depleted.

[2] Anthropocene could be defined as the period during which human activities have had an environmental impact on the Earth regarded as constituting a distinct geological age. The name appeared to indicate that the Holocene period which characterizes the geological period from -12.000 years was registering a significant breakdown. However, among geologists there is no agreement on the exact period it covers, most arguing that the Industrial Revolution is the starting point, while others consider that natural forces and human forces became intertwined much before.

C. Ghymers (✉)
Interdisciplinary Institute for the Relations Between the European Union, Latin America and the Caribbean; Robert Triffin International, UCLouvain,
Brussels & Louvain-la-Neuve, Belgium

© The Author(s) 2024   59
A. Ribeiro Hoffmann et al. (eds.), *Climate Change in Regional Perspective*,
United Nations University Series on Regionalism 27,
https://doi.org/10.1007/978-3-031-49329-4_5

catastrophic "geological proportions" – has been known for many decades, not in geological terms but as the very fact that many human-induced permanent changes to the Earth have been identified early by the oil producers[3] and the US administration.[4] In more recent years, it has become scientifically admitted that these changes are directly threatening humankind and even life on Earth.

Despite the convergence of indicators with indisputable results of many scientific observations, no effective measure has yet been taken, and the policymakers' awareness is even lower than in 1989.[5] While the first important reaction emerged again only in 2015 with the multilateral agreement reached with the Paris Agreement, it has remained mainly on paper and discourses, with few coherent binding targets. Only the EU and the USA have recently enacted targets and tools, but the rest of the world is lagging behind or still disagrees. It is already clear now (in February 2023) that these commitments are definitively out of reach as no significant collective actions have been taken on time despite many discourses and commitments. Therefore, we are obliged to acknowledge a massive failure both of market, governance, and citizens. All actors seem to continue their collective denial refusing to see the unsustainability of the global economic system and the resulting collective suicide, even multilateral institutions (like IMF or European Commission) by showing that a net-zero emissions for 2050 could be easily reached with few economic costs.

Global warming puts radically into question the way of life and production that the overwhelming part of the world has adopted up to now. Therefore, the wait-and-see with the complicity of technological options or econometric models is used to disavow and to escape radical changes. We mean that a genuine solution requires much more than green policy measures or technological progress but rather deeper global changes in the way to measure and conceive production, consumption, and therefore global governance.

The governance failures to solve are directly visible in what we consider as the most important measure of the size of the global disease which is also an indicator of effective (no)policy changes: the underpricing of fossil energies (market failures) worsened by the direct subsidies given to their production or consumption (governance failures). If effective measures had been implemented since the Paris Agreement (or before), these systemic direct and indirect subsidies would have

---

[3] As early as the 1950s, energy companies ordered several scientific studies, all concluding there was a causal link between the use of fossil fuels and global warming, implying therefore a questioning of their future exploitation and profit.

[4] At the beginning of the 1960s, the *Jason Committee* had already established scientifically the threat of global warming for life on Earth, and in 1979, among other official reports, the *Charney Report* predicted scientifically "a warming of between 2 and 3.5 degrees" for 2035, irreversible without total elimination of fossil energies.

[5] Noordwijk Conference that had reached an almost universal consensus for enacting an international compulsory limitation of CO_2 emissions but that failed under vested interest "pressures" on the USA.

started to decline. According to the IMF published data,[6] these subsidies to fossil energies not only did not decline, but rather (direct and indirect subsidies) they have increased from about 5.4% of world GDP in 2015 (Paris Agreement) to 7.1% in 2022. Expressed in current dollar, these total subsidies have been increased since 2015 by a huge $2600 billion, leading to an annual waste of $7000 billion (i.e. $7 trillion), which is double the additional investment necessary each year for decarbonizing the whole planet. These figures are the sum of direct (or explicit) subsidies and implicit ones. Direct subsidies measure the difference between the price paid by the fuel users and the effective financial cost to supply fossil energies. These "cash" subsidies increased from 0.6% of GDP in 2015 to 1.3% in 2022. This amount means that $1326 billion of budgetary resources are wasted for spurring CO_2 emissions, in contradiction with political discourses! Implicit subsidies measure the difference between a fuel's full social cost and the price paid by the fuel user, exclusive of any explicit subsidy. These full social costs include local air pollution, climate change, and broader externalities and amount to 82% of total subsidies ($5710 billion in 2022). The sum of direct subsidies and indirect ones measures the costs imposed on society due to consuming fossil energies at prices lower than their real costs. Any serious economist should consider this amount as negative outputs that should be deducted from world GDP and total factor productivity. This means that each year global GDP is lower by 7%, i.e. the growth rate is in fact very negative!

In the EU, before the subsidies linked to the pandemic, the annual direct subsidies are estimated to reach a total of around $100 bn in 2018 (0.65% of GDP). The EU total subsidies should be around $470 bn or 3% of GDP (European Commission, 2020). In the LAC economies, they are estimated to be 5.1% of GDP. After Saudi Arabia, Venezuela, and Iran, Russia has the highest degree of subsidies, wasting each year around a quarter of its GDP, i.e. $3000 per capita, for damaging the planet.

This indicator is emblematic of the worrying hypocrisy of our denial, since governments that signed the Paris Agreement are behaving in the exact opposite direction to their formal commitments, worsening global warming and exposing their direct, personal, political responsibility to the judgement of history and even to not so far reactions against them.

At the world level, it is obvious that our economic system remains essentially based upon cheap fossil energies which play the role of a key production factor increasing decisively productivity. The Industrial Revolution itself emerged in Great Britain and in Walloon Belgium in the eighteenth century because, among other key factors, energy prices were significantly lower than elsewhere thanks to their abundant coal deposits. This comparative advantage became so important that economists like W. S. Jevons (1866) questioned the sustainability of the economy warning that the British prosperity was too closely dependent upon a non-permanent endowment of cheap coal. His central thesis – which remains true for our time by changing "coal" for "carbon" – was that the British competitiveness over global affairs was transitory, given the finite nature of coal as its primary energy resource. "…without

---

[6] Black, S., Antung L., I. Parry, and N. Vernon, (2023). IMF Fossil Fuel Subsidies Data: 2023 Update

it [coal] we are thrown back into the laborious poverty of early times". He had already raised the systemic question of sustainable development, not in present terms of environment and global warming but in his lucid observation that cheap energy was a key but also the most fragile factor of temporary prosperity, as far as alternative cheap energies don't exist. Jevons' analysis remains valid today on some points. First, his "Jevons Paradox" saying that technological improvements that increased the efficiency of energy use led paradoxically not to reduce the energy problem but will increase consumption of this energy in a wide range of activities. He argued that, contrary to common intuition – and to present ecologist or government recommendations – attempts to reduce energy consumption by increasing energy efficiency would simply raise demand for energy in the economy. For our present situation, this raises two essential questions: the Jevons paradox shows that technological progress increases carbon consumption only in case of letting the free markets transfer productivity into lower carbon price. He didn't conclude that the correct policy is fiscal and financial reforms (increasing the relative price of carbon issuance, imposing higher emission standards, financial risks, and financing research). Jevons recommended cleverly to dedicate part of the benefits of cheap coal to cut public debt and righting social ills by investing in collective goods for creating a more just society: "We must begin to allow that we can do today what we cannot so well do tomorrow…. reducing the burdens of future generations". The contrary of what we do: we accumulate exponential public, social, geopolitical, environmental, and intergenerational debts.

From these basic lessons combined with the frightening picture given by IMF, we immediately can deduct that not only free markets have demonstrated their inability, by definition, to incorporate crucial negative externalities, but most economists and all the authorities of all societies in the world have been also unable to react and most of them continue to deny, either by refusing scientific reports or by believing that technologies and minor adjustment in carbon prices would get rid of the problem, without need for radical changes.

Identifying the Deepest Roots of the Governance Failure in Climate Change

On top of market failures, a democratic failure pushes governments to continue even to misallocate resources and to disincentive alternative clean energies by maintaining or increasing their direct subsidies to burn fossil energies! It means that voluntary policy decisions maintain important and growing flows of public money for selling fossil energies at a retail price which is below the energy's supply cost for nontargeted users.[7] At the world level, these financial direct subsidies reach some $1.3 trillion per year in 2022. This way to channel stupidly scarce public resources

---

[7] Targeting subsidies to vulnerable people is justified but it is generally not the case.

towards the production and consumption for issuing more carbon in the atmosphere has additional economic effects very damaging. In addition to stimulating CO_2 emissions and discouraging innovation in green energies, provoking misallocation of investments and subsidy cost to public finances, they imply that less alternative expenditures (i.e. social and research) could be made or more taxation on the rest of the economy, lowering the growth rate of the economy. Even more shocking, these subsidies counteract directly the few other measures adopted to fight climate change.

This "short-termism" behaviour is common in any government whatever the political or economic regime and whatever the degree of democracy or economic development. This form of denial is facilitated by the high degree of uncertainty on the future effects of carbon, for example, on the non-linearity of global warming and highly probable tipping points provoking collapse of ecological systems. It reveals a universal inability to fulfil the precaution principle and the priority mission of any government: to ensure the production of this vital public good of preserving their societies from global suicide.

Therefore, we identify the most serious issue in this universal disavowal force that seems to be a built-in flaw in human societies to such a point that even democratic order fails to overcome the short-termism bias of policymakers. This bias has probably deeper explanations.

Very recently, the USA and the EU finally showed a higher degree of awareness than elsewhere. The European Commission initiative efforts to organize a beginning of reaction with its "Green Deal" presented in December 2019 and enacted as the EU Climate law in June 2021. This ambitious plan fixed binding targets to become climate-neutral by 2050 and sets the intermediate target of reducing net greenhouse gas emissions by at least 55% by 2030, compared to 1990 levels. Even in this "best case" in terms of objectives, a dangerous degree of procrastination and denial remains present through the confidence in the used models and in the tools (subsidies and Emissions Trading System – ETS). The proposed means seem "too small too late" for facing the high irreversibility risks, particularly as regards the carbon pricing, since the ETF will cover only 40% of the CO_2 emissions (Varga et al., 2022).[8] The transition is not yet recognized as an urgent need for radical changes in financial sector and relative prices[9] with impacts on every economic agent or citizen behaviour.[10] Even the involved economists continue to apply simplistic models. The universality of this irrational procrastination induces to think that the issue is much broader than technical aspects like the underpricing of carbon or to only be due to private firms and vested interests in search for selfish short-term profits. It is the capacity itself of societies to react on time by implementing the required correction

---

[8] This E-QUEST model of DG ECFIN (European Commission) shows that a carbon tax of $100 in 2030 raising to $600 could reach the net-zero in 2050.

[9] Only 16.8% of the answers to the question "how to reduce the present energy consumption" refers to price increases. Online survey *l'Arbre des possibles*, retrieved March 2022. https://www.arbre-despossibles.com/base/sondage_results.php?code=Energie&ipos=1

[10] https://www.sciencedirect.com/science/article/pii/S0264999322001572

policies which is the key problem and which must be solved urgently to prevent the eradication of our societies.

We identify the roots of our unsustainability in our materialist rationality (Ghymers, 2021a),[11] generating a male predatory (Ghymers & Gonzalez Carrasco, 2016),[12] unfair attitude with nature that used to be necessary for our past survival and material development. The result is an abusive rationality of the male predator which is exposed to be eventually castigated by the self-destruction of its unnatural material results, turning the expected material benefits and rationality into negative output, i.e. showing the irrational male behaviour.

The best example is the fact that economic growth and its measure in terms of GDP have become a pathetic illusion as far as the negative output of CO_2 emissions and other depletions of natural species and resources are not duly deducted from the national accounts. An indicator of the gravity of the popular illusion is given by the majority belief (56.7%) that technical and scientific progress alone would solve the energy problem.[13] Another example of this paradoxical blindness: each year of delaying resilience-enhancing policies in infrastructure sectors could also cost an additional $100 bn in avoidable disaster impacts in LDCs. Not investing in decarbonization will cost even more in the long term than acting strongly now but also because there are significant opportunities for investors, workers, and consumers. According to the World Bank research, investment of $1, on average, yields US$4 in benefits. For example, replacing the costliest 500 gigawatts of coal capacity with even cheaper solar and wind would cut annual costs by up to $23 bn per year and yield a stimulus worth $940 bn, or around 1% of global gross domestic product. A shift to low-carbon, resilient economies could create over 65 million net new jobs globally to 2030.

How to explain that rationality has led us to this collective irrational behaviour of generalized disavowal making citizens unable to see that short-term costs would prevent catastrophic costs later? The answer is probably not only a "political economy" one but, following Sébastien Bohler (2019), rather a neuron-chemical one due

---

[11] The Systemic Nature of the Global Crisis and Some Principles for Tackling it. In B. Guilherme and others (Eds), *Financial Crisis Management and Democracy*. Springer. Neuroscience could explain that our brain structure has pushed the world towards a male rationality based upon a binary thinking that has led to a dichotomist conception of life. The spectacular material results have produced a terrific illusionary bias by separating our perception of our power from our impacts on life and our planetary systems, which in turn has created an "illusory commodified reality" that is engaging humankind in a dehumanized deadlock.

[12] Can predominant intercultural management dimensions based on the binary opposition of cultural differences bridge the cultural divide between CELAC and Europe? COLLOQUE IDA – CERALE- Centre d'Etudes et de Recherches Amérique Latine – Europe, ESCP Europe Business School, Paris. We are convinced that gender disequilibrium is one of the main factors of the unprecedented destruction of our planet and the separation between humans and nature. On this see the work of Francisco Varela and the Santiago School of Cognition. See also Myhre, S. (2019) Cambio global, mujer y academia. Paper presented at *Congreso Futuro*, Pontificia Universidad Católica de Chile, Santiago, January 2019.

[13] Web survey *L'Arbre des Possibles*, op. cit.

to the successive composition of the human brain. The rationality results from the countervailing power of our cortex over our animal *striatum* which is dominated neuron-chemically by our animal nature through the issuance of dopamine. The invasion of materialism combined with a lack of growth has weakened the collective ability to plan for longer term. The popular reflex of fear gives preference to the immediacy, "democratically" refusing any longer-term considerations, rejecting the elites in power viewed as necessarily guilty for the current deteriorations, and, opting for immediate advantages, condemning democracy to populism. The economic consequence of this "irrational destructive behaviour" could be translated in a global "tragedy of commons" where the essential common public goods is paradoxically depleted by individual materialism impeding to perceive the exponential costs of inaction compared to the higher benefits from short-term adaptation costs and forward-looking investments. It is an effective democracy failure which does indeed worsen the market failures as a result from populist slogans and from complex manipulations by vested interests (lobbies, financial powers, corporatist groups, and "rent-seekers"), since our political democracies are excessively biased by a lack of economic democracy. This defect is even more perverse in autocratic or centralist state control.

Identifying the Coherent Set of Global Policy Tools to Front-Load

Due to the global "dictatorship of materialism", the only effective recipe for stopping on time the destructive mechanisms of global warming is to act on the same level as the materialism inherent in our cultures due to the disavowal led by the dominant animal part of our brain.

This "recipe" consists of correcting the most important materialist incentives for issuing CO_2. Indeed, it is urgent and powerful to scrap subsidies to fossil energies and to transfer a part to lower-income households and to clean alternative energies combined to a rising floor for carbon price or carbon taxes (Schulmeister, 2021; Parry et al., 2021), up to what could be estimated as their efficient level taking on board as most as measurable their negative spillovers.[14] This could provide much more resources than necessary for a total decarbonization of the world ($7 trillion per year). Econometric models (Parry et al., 2021) show that an optimal relative price for carbon would lead automatically to a net-zero goal for 2050. We see in these models a more dangerous risk of disavowal by creating the illusion that transition would be automatic and almost costless. Most economists use "Integrated Assessment Models -IAMs" which conclude that raising moderately the price of carbon, for example, to $50 per ton (Parry et al., 2021) or/and using subsidies to

---

[14] For instance, on the base of IMF data base on IMF Fossil Fuel Subsidies, https://climatedata. imf.org/

alternative energies (Varga et al., 2022) would produce decarbonization, believing that market will take care of it all. The shortcoming flaws of these models are analysed in several technical papers which all indicate the high sensitiveness of their results to methodological aspects[15] making them unreliable for facing the risks of irreversible catastrophic scenarios. Their dominant acceptance is a clear demonstration of the overwhelming disavowal illusion.

On the contrary, we sustain that the solution is not only an urgent increase in relative prices but to correct a set of related systemic dysfunctionalities of the world economy. An increase of the investment rate must be effective not only in advanced economies but in all others too. This requires huge capital flows from advanced economies to the others, i.e. a significant decrease in the consumption share in advanced economies GDP.

We sustain that this couldn't result without correcting two other more complex *price distortions* which are also global and need urgent multilateral agreement:

- First, the too high financial return with respect to real economy profitability.
- Second, the too low yield on safe assets in the key currency, the dollar, which indicates an inadequacy of the International Monetary System (IMS) to fill the financial gap for decarbonization.

The combination of these two price distortions explains that the issuer of the key currency not only absorbs the global net saving but overall impedes to rebalance the gross capital flows towards decarbonization in the emerging and developing economies in accordance with the Paris Agreement. In fact, this is merely the manifestation of the Triffin Dilemma (TD) (Ghymers, 2022).

The other side of the coin of the TD is generally forgotten: the "built-in destabilizer" (Triffin, 1959) which shows the permanency of the Dilemma through the pro-cyclical variations in global liquidity (GL) and the impossibility for the Fed to counteract it. This means that, even more than before, the TD explains the damaging spillovers of financial instability with perverse reversible capital flows in Emerging and Developing economies (EMDEs). This impedes to fill the financial gap for their decarbonization, as far as the dollar remains the basis of the IMS because it is the determinant of the GL. The TD explains most of the difficulties to finance the decarbonization in EEs-LDCS because (i) the saving flows are biased towards US overconsumption, and (ii) mostly gross international capital flows determine LDCs investments, being in dollar and reversible, penalizing decarbonization in these economies.

Mark Carney estimates that about at least $130 trillion are needed for reaching zero emission for 2050.[16] This implies to dedicate each year at least 4% of global GDP only for energy transition investments. For the low-and-median-income LDCs (LICs and MICs), the World Bank estimates that their climate transition needs

---

[15] The best analysis is Stern et al. (2022).

[16] https://www.forbes.com/sites/jillbaker/2021/11/08/mark-carneys-ambitious-130-trillion-glasgow-financial-alliance-for-net-zero/

would require increasing inflows from \$425 billion per year (period 2019/21) to \$1.7 trillion by 2030. For all LDCs, the need for climate investment flows per year by 2030 could reach \$3.4 trillion (World Bank, 2022). Based on this assessment, climate action in developing countries presents a gigantic and growing financial gap (World Bank, 2022). As shown in Fig. 1, present situation indicates this gap.

Therefore, two inevitable systemic changes must go along with the carbon price corrections in order to rebalance gross financing flows towards the required real investments in green energies (4% per year of global GDP up to 2050):

- The IMS must move to a single ultimate liquidity which is not any more the debt of an economy but is issued by a multilateral central bank acting as the Global Lender of Last Resort (GLOLR). A regulation of GL as a public good would become feasible. In fact, both do already exist by upgrading the IMF into a Global LOLR issuing the most stable reserve – the Special Drawing Right (SDR)[17] – against national reserves. With such a superior stable safe asset, the price distortions of both the safe asset yields, and the financial yields could be reduced. In other publications we explained in detail these mechanisms of the growing pro-cyclical GL (Ghymers, 2021b, 2022).
- As recommended by the Network for Greening the Financial System,[18] the macroprudential regulation must include the nature-related and climate risks in all financial assets because they impact not only on the stability of the financial system, but their underestimation corresponds to a subside to carbon production and ecological damages. This implies the need to include these concerns in the

Fig. 1 Total government clean energy investment support enacted since the start of the COVID-19 crisis, by region: \$1215 bn. (Reproduced from **IEA Government Energy Spending Tracker** Government Energy Spending Tracker – Government Energy Spending Tracker – Analysis (IEA))

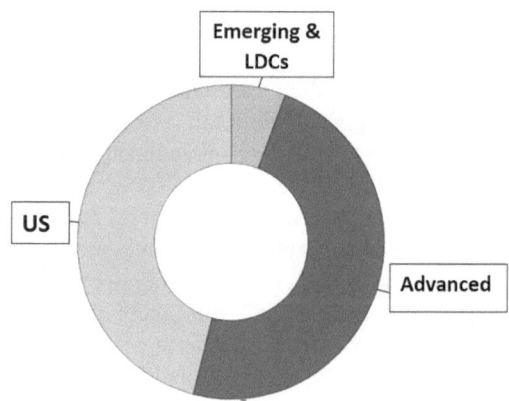

---

[17] Special Drawing Right is a basket of the five main reserve currencies, created in 1969 for being the official international reserve. See SDR working group, Robert Triffin International (RTI) (2015). Using the Special Drawing Rights as a lever to reform the International Monetary System. *International Monetary Issues n°2*, Louvain-la-Neuve, Versant Sud.

[18] The NGFS, launched in 2017, is a group of central banks and supervisors, which on a voluntary basis are willing to contribute to the development of environment and climate risk management in the financial sector and to mobilize mainstream finance to support the transition towards a sustainable economy. It gathers already 121 members and 19 observers, covering almost the world and all the systemic banks.

mandate of all central banks. Indeed, all the monetary tools as collaterals, bank reserves, and other ratios should incorporate differentiation in the asset values in function of the sustainability risks. Furthermore, the financial sector plays a key role in the financing of investments in low-carbon or decarbonization activities. But this is not enough; the high uncertainties on future technologies and prices should be compensated by giving immediately guarantee of lower capital costs and higher yields to new low-carbon equipment or activities. Financial guarantees and – as proposed by M. Aglietta (Aglietta et al., 2022) – the creation of a "carbon asset" making "cash-able" or usable the social value of avoided carbon could accelerate the decarbonization of the present output structure in a modular form in each country. For example, the risk of funding would be reduced by accepting these assets in the eligible collaterals at the central bank or by their securitization in "green bonds".

A coherent set of policies requires thus a triple correction of deep price distortions at global level that implies significant systemic changes in behaviours and governance, something which is not properly acknowledged by citizens and feared by all governments.

Box: The New Form of Triffin Dilemma (TD) Explains the Global Financial Spillovers and the Pro-cyclical Global Liquidity

Triffin dilemma (TD) is the incompatibility for a national currency to be the best international liquid standard (safe asset) because its central bank cannot simultaneously regulate domestic and international liquidity. In the two last decades, the wholesale monetary markets have essentially relied upon collaterals, most of them in dollar-safe assets. But the Fed and the US Treasury cannot supply sufficient safe assets for matching their global need. The exponential expansion of the unregulated nonbank sector, joined to the massive use of the dollar by the emerging economies, must face an insufficient supply of dollar safe assets. Therefore, the "repo markets" manufacture pseudo-safe assets (out of regulation and access to the Fed) by their own intermediation but with a higher-risk leverage, by successive securitization with "rehypothecation" of US T-bills and with use of nondollar safe assets for making loans to riskier debtors. This is an endogenous creation of monetary basis on the repos for expanding global liquidity (GL). However, once a recession or liquidity crunch is expected, this expansion is reversible, proving the systemic incoherence of relying upon a national currency for international purposes. Since the dollar is endowed with a higher "moneyness" – a liquidity monopoly – it introduces a differentiation among the collaterals composing the liquidity basis according to their effective degree of "moneyness" (their resilient liquidity): the "external" collaterals, which are exogenous policy tools issued by authorities (central bank and Treasury, mostly from the USA), and the "internal" collaterals, which are mainly pseudo safe assets. This discrimination creates a cyclical "dash-for-cash" in dollar-safe assets, something like a "Gresham law" inside the "shadow monetary base" leading to a cut in the basis of the reverse pyramid of GL shadow monetary base, with a multiplied restrictive effect on private GL. This is the main cause of the GL pro-cyclical behaviour which provokes perverse spillovers on the capital movements to LDCs. This instability is inherent to a key currency once it is unable to issue enough liquid debt in facing the huge gross flows proportionally based upon more collaterals than before, due to the shift in liquidity sources from banks to nonbanks while the US economy is relatively in contraction with respect to the global financial needs. TD explains the cyclical scarcity of dollar-safe assets which makes instable and reversible the private global monetary base, the GL, and the capital movements, something which could disappear with a multilateral reserve currency like the SDR, which is not the debt of a national economy, does not impose neither financial nor political costs to the participants, respects policy autonomy, and makes feasible the move to a multi-polar monetary system. On the contrary, the emergence of competing key currencies (like the Chinese Yuan) without a common, single liquidity standard would create even more unstable due to the impossibility of reaching an equilibrium exchange rate between two competing dominant currencies (Kareken & Wallace, 1981). Therefore, the principle itself of key currency is functionally condemned to disappear.

The Framework of the Bi-regional Strategic Alliance for Meeting Climate Change Challenge

The General Disavowal Results from a Typical Prisoner's Dilemma Which Inhibits Energy Transition

Redeeming the intertwined trilogy of failures – markets failures, public failures, and democracy failures – would require too deeply rooted changes in a too short time in a global context not favourable to cooperation: growing antagonisms in the geopolitical context, weakening of multilateral institutions, rise of populism, "post-truth", and reject of scientific elites. The irreversibility of climate change and the consequent degree of high urgency require to tackle pragmatically the procrastination of policymakers that maintains the three cumulated failures. The authorities face domestically the general short-termism under national populist pressures and externally the risk of free riding and the difficult burden-sharing among nations or regions. This status quo represents a typical prisoner's dilemma which inhibits national authorities: collective irrational losses result from opposite "rationality" between decentralized policymakers in competition when each is issuing spillovers to the others. Indeed, under uncertainty about the "real model" for climate and in the presence of mutual spillovers, the lack of trust both at national and international levels explains the common disavowal and the lack of cooperation. Domestic opposite political parties and conflicts between countries combine for impeding faster actions by lack of consensus. Furthermore, remaining in power are biased national policies towards short-term and noncooperative behaviours with other countries.

What Are the Basic Principles for Getting Out of This Prisoner's Dilemma?

- ### Building Trust Among National Policymakers at Regional/Bi-regional Levels

First, the logic for finding a concrete solution is to build mutual trust among countries by using the common threat and costs of climate change for setting up systematic exchanges among peers at regional and interregional levels, for dealing freely on the common threat, the possible options, and the common goals. This simple method to spur cooperation on energy transition consists in making more tangible the win-win game of regional/bi-regional cooperation first among national technicians and policymakers themselves and then for public opinions.

- ### A Two-Tier Regional/Bi-regional Scheme

At first glance, this could appear as much about the same as what already do the policymakers. But the second principle is to proceed in a successive regional/bi-regional two-tier scheme: the basic step consists in creating a regional/bi-regional peer network (closed doors and Chatham rule) among the national technicians (civil servants and experts in energy transition) with a personal mandate for collegial monitoring of the respective national issues and strategies, while making explicit that each peer does not represent his minister, country, or region but is only mandated to freely exchange and speak on a purely technical and personal basis. These free dialogues allow for triggering a two-tier group dynamics at regional and bi-regional levels with a group momentum questioning the collective irrationality and pressure for a more cooperative approach.

The decision-making step is the conventional one which exists at ministerial or Head of State levels but is actively fed by the momentum created at the basic technical levels.

- ### The Need for a Bi-regional Approach on Energy Transition

The multilateral level, theoretically the optimum one, is not presently the most practicable. Indeed, the kind of feasible coordination remains very weak, and the present geopolitical context doesn't give much illusion of spontaneous progress in the foreseeable horizon. Therefore, the intermediate option of regional coordination makes sense not only among member states of a region but also for easing the cooperation or coordination between regional blocks which have been already working in looking for common economic interests and social goals.

We sustain the idea that the bi-regional level could be much more than a second best because it creates a deep catalytic effect on each regional cooperation for making possible to agree upon common interests when preparing the meetings for bi-regional exchanges. The idea is to use the cooperation with other regions for accelerating each own region cooperation and their specific momentum with positive effects upon public opinion awareness. No formal change and no new procedure are required and the tools do exist. Our proposal is just to bet on activating the existing endogenous win-win game first between EU and CELAC regions, the two most like-minded areas. This bi-regional dimension provides concrete steps towards a workable dynamic of energy changes inside the existing system through a priority focus on cooperative tools for energy transition. We sustain, by experiences, that the bi-regional cooperation could be – under certain conditions – an efficient way to win time and synergies in breaking the prisoner dilemma obstacles to an effective energy transition. The detailed mechanisms and method are explained in section below.

Furthermore, this proposal would permit to give an effective content to the 1999 objective of a bi-regional strategic alliance the EU and the ALC regions have been pretending to build during almost 25 years but which is still missing. The proposal is to quickly trigger a new dynamics by giving a technical content to the energy transition that both regions are condemned, anyway, to organize in the urgency and to bargain at multilateral level. The emergency aspects and the gravity of the challenges to meet in a more complex geopolitical environment should provide new pressure on actors from both regions for looking for common strategic interests in the existing bi-regional cooperation among like-minded partners. Therefore, a single key content is provided to the strategic alliance that is the legitimate purpose of the existing bi-regional summit. Both regions share the same kind of values and society; they share common interests. Both regions could win competitiveness as well as power vis-à-vis other third powers or economic blocks. Anyway, their coordination or consensus would give them a more protagonist role in global economy and in multilateral organizations. The positive results of this kind of bilateral win-win game would help to attract the interest of other regions for extending the scope of the coordination to them, easing the process towards the ideal multilateral implementation.

How Could These Principles Give Better Results?

Our proposal includes a new, specific method based upon solid experimentations and personal experiences in various contexts (EU, Eastern and Central Europe, Africa, and Latin America) (Ghymers, 2005) where all were affected by similar prisoner's dilemmas in economic policy coordination, i.e. a domain where policies are national but with spillovers upon their partners.

The bi-regional summit would implement these principles to launch a cooperative momentum with the priority focus upon the measures to reach on time a net-zero emission in both regions and, if possible, to extend the method to third regions, becoming so multilateral. The experiments show that a bi-regional dialogue in a two-tier scheme, reproduced in two steps at regional and bi-regional levels, speeds up trust and consensus on right policies.

A Two-Tier-Two-Step Dialogue

The two-step scheme: first step among technical peers, organized as two separate regional dialogues in charge of preparing the collegial monitoring of the respective national plans for energy transition: these technical groups form only consultative regional committees for improving the quality of information and for preparing the second step of the scheme. Ministerial dialogues are the respective regional/bi-regional decision bodies on energy transition. The two-tier dialogue: first level only in each region working separately among member states successively through the two steps (technician committee first, feeding Ministerial Council); second level when both two-step schemes (technician committee first preparing Ministerial Council) meet successively at bi-regional level. This two-step-two-tier-dialogue generates a powerful game dynamic which accelerates the degree of awareness and implementation of the best policies, as a result, not only from exchanges of best practices but also from the emulation among peers combined to the pressure of a mutual scrutiny which creates some kind of permanent "check-and-balances", issuing collegial advices/criticisms on what is feasible with respect to what is done. The technical peers benefit from a valued role in their own country which provides a strong individual incentive for contributing.

The regional cohesion is never spontaneous and does not depend so much upon institutional development, except the minimal two-tier scheme proposed which creates the accelerating momentum. Cohesion results from the gradual development of personal contacts and collaborative efforts between experts of a region's countries, which is progressively transferred to Ministers and encourages a common culture. A basic regional consensus emerges and creates a climate of collegial trust. Since the prisoner's dilemma is the major obstacle to regional coordination and hinges on uncertainty about other players' behaviour, increasing technical communication among them and asking them to issue collegial opinions clearly improve their chances of finding a way out of the regional suboptimal situation of a lack of cooperation.

This dynamic process is enhanced by the effect of the bi-regional dimension, where the same momentum should appear across European and Latin experts as far as participation is ensured with some continuity by the same persons. Also, this is not theory but experiments as well as socio-psychology observation of concrete organizations.[19] The duplication of the two-step method at the bi-regional level

---

[19] We refer here not only to specific experiences in regional or bi-regional meetings but to the general group dynamics observable when any organized group characterized by some continuity must bargain with another one endowed also with some continuity something which concerns both parties.

brings an additional incentive for the participating experts who benefit from the other region exchanges, monitoring, and questioning. Furthermore, it capitalizes upon the regional dimension in each regional entity (EU and CELAC) which is stimulated by the need to "speak with one voice" in the bi-regional dialogue for identifying common interests as well as making clear the divergences to overcome together with technical as well as political incentives to reach bi-regional consensus. The experiences in observing bi-regional and interregional other negotiations both in Europe and in other regions of the world make clear that much more regional cohesion results when one block must prepare a bargaining session with an external partner or block.

The Key Role of the Technical Committees

The distinction between the expert networks (committees) and the Ministerial Council is very important for creating the group dynamics based on a bottom-up process of free exchanges stimulating personal initiatives. The committees hold a collegial scrutiny inside an effective monitoring of existing and possible policy measures but without taking any decision. On this basis more weight are given to the experts, so spurring the consensus by informing the Ministers on how to make the national policies coherent and converging to the net-zero goals.

Paradoxically, in this dynamic dialogue, the most important role belongs to this consultative, non-decisional committee. It generates reciprocal confidence among participants which permits to get genuine exchanges of information, freer critical opinions, and creative debates on the most urgent measures to propose to their own Ministers. National experts obtain some key information on the partners thinking, on alternative scenarios, and on concrete basis for proposing concrete realistic elements to their Ministers in view of the bargaining among national and regional authorities. Its fundamental role is to generate trust among national experts with good communication flowing reciprocally among them. This reciprocal confidence and the fact to be in a continuous game eradicate the cause of the prisoner dilemma, creating the conditions for deeper and faster cooperation among states and regions. By building a regional consensus in both regions, a stronger coalition EU-CELAC on priorities for energy transition could result.

The Dynamic Collegial Monitoring for Breaking the Prisoner's Dilemma

The concrete working of this *two-step-two-tier* method for breaking the prisoner's dilemma relies upon the principle of a collegial monitoring of energy transition plans among peers with quantitative objectives set in a harmonized form to be fully comparable. Peers assess reciprocally each plan and discuss the hypothesis, existing measures, and possible policies to propose at regional level before opening dialogue and reciprocal monitoring at bi-regional level. Due to the similarities between their functions, difficulties, and responsibilities, the technical peers generate a collegial spirit among them. The result is a better understanding of each national position and constraints to solve together. Group cohesion becomes an efficient tool for innovation and coordination. This *bottom-up* approach should create a powerful endogenous dynamics of technical cooperation, which would directly help national policymakers to identify better priorities, issues, and pragmatic solutions, while saving conflicts and resources and increasing ownership of the policies. In turn, this collective value added gives to the technicians a higher motivation, a specific role vis-à-vis their Minister and therefore a better leverage inside the national decision-making process beneficial to their country and to the region. It results a better governance and a higher awareness of the challenges to meet. Another advantage of this specific type of technical network is the longest continuity of the technical teams, if compared to the politically responsible individuals or officers directly linked to changing minister cabinets. This permits to maintain the cooperation on technical aspects whatever the political cycles and their conflicts.

Referring to the analytic figure of the prisoner's dilemma, the challenge consists in increasing communication for making tangible the dynamic benefits that each participant expects to obtain by taking advantage of the regional monitoring and its value added for each of the partners. The basic channel to grant incentives is the process of credibility generation which allows for reaping early the full benefits of peer pressures towards sustainability. This credibility, both for the technicians and their country, can be provided from a regional monitoring on national net-zero transition policies with rules, solid control schemes, and financial incentives which must offer individual advantages to its participants. Credibility of engaging on sustainability path influences directly investments, savings, and financial assistance, with impacts on growth, by reducing the uncertainty. Mutual regional monitoring of energy policies generates also a self-gratifying mutual knowledge, a better communication, and a greater confidence among autonomous participants. This first step is necessary to reduce the uncertainty regarding both the potential benefits and the mutual confidence.

Cooperation thus becomes a directly useful instrument for national decision-makers by enhancing the credibility of their policies and directly helping them carry out their own political commitments as well as better financing modalities. Cooperation is based on self-interest tempered by the better knowledge of the partner's interests and difficulties that could generate for them both negative and positive feedbacks. To use game-theory terminology, the probability increases that all games will have dominant strategies that coincide with the net-zero measures at regional level that fit with the social optimum. We name this a "competitive cooperation" (Ghymers, 2005) process.

In particular, the *three important components* for the efficiency of the monitoring scheme for energy transition are the following:

- First, the use of numerical benchmarks with precise time schedule and procedural commitments in case of deviations, able to attract public and financial market attention and trigger political incentives especially for countries with financing gap depending upon regional or multilateral lending.

- Second, to include a link between the degree of implantation of the package and the degree of regional or bi-regional support to get access to financial resources for their energy transition. Each region would issue first its collegial assessment based upon the expert technical reports, and then this region submits it to the other region for discussion and conclusion at the bi-regional level for obtaining more credibility.
- Third, to bargain with multilateral and regional financial institutions for grouping the financial resources into a specific bi-regional fund for energy transition cooperation submitted to criteria assessed by the bi-regional two-tier scheme.

The experts must feed their respective Ministers before the Ministerial Council with these results which open the scope of the obstacles and feasible options to discuss in the Council. This method presents the advantage to create incentives for upgrading governance as a result of the principle that "the less credible the initial national transition policy is, the greater the net potential benefits of the collegial monitoring result" for any country participating to the monitoring game with its regional peers, and the more powerful becomes the "group dynamics" for the benefits of all. Participants compete in strengthening their respective roles, their managing capacity, and their knowledge to show the professionalism of their respective administrations in charge of energy transition. This induces to a process of *competition to improve* and of creative initiatives for being cooperative. This is the competitive cooperation among peers for trying to demonstrate the qualities or defects of the policies of their partners as well as for their own country, but they also learn to respect the respective national or regional peculiarities. The same dynamic scrutiny works as regards the identification of common interests, priorities, and about the strategy they recommend for making it possible to speak with a single regional voice, when they meet their partner's colleagues from the other region. The fast improvement in mutual and personal knowledge allows for building trust among participants leading to reduce the obstacles that hamper cooperation among sovereign authorities, i.e. solving progressively the prisoner's dilemma through intense communication.

With the use of new information, important tasks can be implemented that could not have been planned – not even at national level – if they would have to be channelled through diplomats in official mandate before being able to hold formal regional negotiations. Therefore, sharing the analysis and the information about difficulties stimulates the gradual creation of an emulative *team spirit* among energy, environment, and economic experts. This group identity acts as a catalyst on cooperation among participants and ministries, because the group helps participants in their daily tasks. Hence, technicians get more credibility and efficacy for preventing conflicts or big policy mistakes. The final receivers of these benefits are the authorities (Ministers) who profit from the information received as well as from the motivation of their experts, who, in turn, experience a personal growth through their role in the regional group. This leads to a dynamic process of convergence *as governance in each country improves, so does regional consensu*s.

A tremendous improvement will be observed in the quality of the information each minister benefits before making decisions, and since this information is also reported to the other Ministers, their council meetings win efficiency and more probability to take on board common interests and spillovers.

Confidentiality, Technical Exchanges, and Democracy

The role of Ministers is to defend constitutionally the national-political positions imposed by the legitimate national sovereignty. The role of technicians and economists is to show them when and which conditions regional/bi-regional cooperation could converge with their own interests or goals. As developed, this role needs first to build trust among them. Ministers always speak as official representatives of their governments. On the contrary, in the technicians' committee, experts speak on their own, never in the name of their hierarchy or country because they don't decide. For creating confidence and cohesion in the committee, the exchanges must remain informal, i.e. made on personal basis; therefore some degrees of *confidentiality* are the inescapable rule for breaking the prisoner's dilemma. This degree of confidentiality is the starting point of any realistic cooperative work; this is why it is applied in many other fields in democratic countries (justice, central bank, governments, political party, professional corporations, scientific experts, etc.).

As regards the democratic aspects, some criticisms could consider the "confidentiality" of the technical exchanges as a nontransparent defect. The answer is simple: without some degrees of confidentiality in the preparation of any decision, participants never would dare to speak freely knowing that their words could appear in the press, impeding any genuine dialogue, and in this case the committee would be useless policymakers would remain block in the status quo. This criticism expresses an ideological confusion between technical consultations and democratic public decisions, implicitly believing that direct democracy could decide on any field.

Furthermore, the kind of confidentiality is merely applying the "Chatham rule"[20] in the sense that the content of the debates could be communicated in a final report but not the identities of the expressed positions, except to their own respective Ministers. Therefore, the results of their exchanges and recommendations – which are not at all decisions – could be made public after the Council of Ministers and without other censorship than the technical and scientific rigour decided by the peers, as in any public democratic debate. Ideally, a public debate should be organized by the expert committee by publishing its synthetic report and its consensual recommendations as well as the eventual divergences manifested among the peers (without naming them). The public debate could be organized – at least among professional experts and specialized media – for ensuring the ownership or reactions,

---

[20] This rule is currently used around the world to encourage inclusive and open dialogue in meetings because it helps to bring people together, break down barriers, generate ideas, and facilitate to agree solutions.

allowing for comparisons with what the Council enacted on the same technical basis. This would introduce the best democratic pressure on the whole process by joining technicalities to value judgments, ensuring broader scrutiny and ownership of technical policy measures. According to our personal experiences, the progress in integration in the European Union was entirely dependent upon the confidential exchanges inside the Monetary Committee (Ghymers, 2005), although this committee was only preparatory of minister councils without any decision role. The same personal experiments occurred in the CEPAL with REDIMA (Ghymers, 2005) for Latin America as well as in the two CFA African monetary unions.

Addressing the Systemic Issue of the Global Financing of the Net-Zero Strategy

The proposed method would be especially helpful for dealing with the two systemic issues blocked by the prisoner's dilemma and its denial expressions:

- The need to see the short-term costs of decarbonization as a highly productive investment for longer-term social and financial returns for all.
- The enormous financing gap in LDCs is an obstacle to decarbonization. In addition, in LAC region, several countries still count on the financial receipts of their oil resources.

It would be crucial to benefit from frank exchanges at bi-regional level on the conditions for fulfilling this gap. All the more that it is not just a mere financial problem but is one key facet of the trilogy of failures mentioned (market failures, governance failures, and democratic failures) which is the result of the incoherence of the global economic system: the principle of the "tragedy of commons" (i.e. those who benefit from causing damages to public goods do not pay for it), creating a systemic divergence between private and social returns. This is clearly the case for global warming resulting from overly low relative prices of fossil energies. This is also the case for the trend towards economic stagnation resulting from global governance failure allowing for overly high relative yields for short-term financing activities with respect to real production investments (second relative price distortion). Furthermore, this is the case for the bias of savings flows towards the US overconsumption against LDCs' investment needs through the overly high relative prices for safe assets in dollar (third distortion: too low relative yields). These three distortions in relative market prices explain dysfunctional behaviours which make mutually reinforcing the microeconomic, macroeconomic, and financial flows: they feed global warming by diverting financial flows from where they would give the highest yields for the common goods. Although these macro-financial aspects do also affect other regions, they are especially crucial for LAC economies, and the like-minded characters with the EU region should make it easier and more powerful to speak with a single voice in global forums, providing also opportunity for third regions to join.

This key issue of the link between green energy investments and the IMS cannot be solved by the EU or the LAC. Nevertheless, working at bi-regional level on possible remedies to this asymmetric IMS would ease the multilateral debates for deeper reforms. For instance, issuing an EU-CELAC joint-communiqué with consensual proposals for developing the role of SDR in conformity with IMF statuses might attract the other regions to reinforce the pressures for IMS/IMF reforms (Robert Triffin Intenrational, 2015). Anyway, without tackling the issue of the Triffin Dilemma, the Paris Agreement disposition on the necessary flows of funds towards emerging and LDCs economies could not be implemented, and financial flows would remain well below the need for net-zero emission on time.

Conclusion

In the present context of weakening multilateralism, the regional level of international politics and comparative regionalism allows for proposing to use more actively the bi-regional level between the EU and the CELAC for making decarbonization a priority for their strategic partnership. The rising gap between the urgency to cut CO_2 emissions and the modest reactions manifests a worrying global governance problem due to a general disavowal in public opinion and policymakers' procrastination. We explain this universal disavowal in modern societies, whatever the political regime and the region, as a prisoner's dilemma which results from a combination of deep neuron-biological features biasing towards short-term views when uncertainties are worryingly increasing, with a democratic failure allowing for power manipulations by vested interests, unfair income distribution, and populist materialism. Global warming is emblematic of the unsustainability of our economic system due to a systemic divergence between private and social returns, mutually reinforcing the microeconomic, macroeconomic, and financial flaws.

The only operational solution to these combined flaws in our economic system in deteriorated geopolitical and multilateral context is to act at the same materialistic level for imposing a convergence between private and social economic returns, in priority by enacting a strong front-loaded package based upon significant increases in carbon prices with social compensatory allocations and multilateral cooperation for moving the IMS to single multilateral reserve allowing to regulate and stabilize GL, together with financial innovations and regulations for increasing the return of decarbonization. The inescapable short-term costs will be rapidly compensated by the benefits of sustainability and, overall, by reducing or preventing the gigantic costs that carbonization is about to provoke soon.

For reaching these necessary changes, our thesis is that the bi-regional cooperation framework between two like-minded regions – the EU and the CELAC – provides concrete ways to make possible effective actions. The necessary condition is to reverse the traditional method used (at regional as well as bi-regional levels) from a "top-down" (political decisions first, technical consultations after) to a "bottom-up" (technical experts free exchanges first providing proposals to policymakers).

This proposed method could give the missing operational content to the bi-regional strategic alliance by creating a collegial monitoring of energy transition. Based on simple sociological mechanisms – extracted from past experiences – our method bets on the powerful group dynamics created among national civil servants and experts organized in protected networks (Chatham rule), which could spur policy-makers' awareness and ease their decisions.

References

Aglietta, M., Bai, G., & Macaire, C. (2022). *La course à la suprématie monétaire mondiale*. Odile Jacob.

Black, S., Antung L., Parry, I., & Vernon, N. (2023). *IMF fossil fuel subsidies data: 2023 update* (Working paper, IMF, 23/169), Washington, DC.

Bohler, S. (2019). *Le Bug Humain: pourquoi notre cerveau nous pousse à détruire la planète et comment l'en empêcher*. Robert Laffont.

Crutzen, P. J., & Stoermer, E. F. (2000, May). The 'Anthropocene.' *IGBP Newsletter, 41*, 17–18. Available from: http://www.igbp.net/download/18.316f18321323470177580001401/1376383088452/NL41.pdf3

Dubos, R., & Ward, B. (1972). *Only one earth*. Penguin.

European Commission. (2020). Study on energy costs, taxes and the impact of government interventions on investments in the energy sector, Trinomics B.V., DG ENERGY.

Ghymers, C. (2005). *Fostering economic coordination in Latin America: The REDIMA approach to escaping the prisoner's dilemma*. United Nations Publication, Book n°82.

Ghymers, C. (2021a). The systemic nature of the global crisis and some principles for tackling it. In B. De Souza Guilherme, C. Ghymers, S. Griffith-Jones, & A. Ribeiro Hoffmann (Eds.), *Financial crisis management and democracy: Lessons from Europe and Latin America*. Springer Nature.

Ghymers, C. (2021b). *The systemic instability of ballooning Global Liquidity as a symptom of the worsening of the Triffin Dilemma* (RTI paper no. 15). Centro Studi sul Federalismo, Torino.

Ghymers, C. (2022). *The new form of the Triffin Dilemma and its built-in destabilizer: The process of relative shortage of dollar safe assets* (IRELAC working paper, 7), Brussels.

Jevons, W. S. (1866). *The coal question; an inquiry concerning the progress of the nation, and the probable exhaustion of our coal mines* (2nd ed.). Macmillan & C°.

Ghymers, C. & Gonzalez Carrasco, C. (2016) Intercultural Management: Bridging the Cultural Divide – CELAC – Europa, in Management interculturel et affinités électives, Europe - Amérique latine et Caraïbe, by CERALE-Centre d'Etudes et de Recherches Amérique Latine - Europe, et ESCP Europe Business School, Paris.

Kareken, J., & Wallace, N. (1981). On the indeterminacy of equilibrium exchange rates. *The Quarterly Journal of Economics, 96*(2), 207–222. https://doi.org/10.2307/1882388

Parry, I., Black, S., & Roaf, J. (2021). *Proposal for an international carbon Price floor among large emitters* (Staff Climate notes). IMF.

Schulmeister, S. (2021). Financial instability, climate change and the "digital colonization" of Europe: Some unconventional proposals. In B. De Souza Guilherme, C. Ghymers, S. Griffith-Jones, & A. Ribeiro Hoffmann (Eds.), *Financial crisis management and democracy: Lessons from Europe and Latin America*. Springer Nature.

Robert Triffin International – RTI. (2015). *Using the Special Drawing Rights (SDR) as a lever to reform the international monetary system*, Louvain-la-Neuve & Torino.

Stern, N., Stiglitz, J., & Taylor, C. (2022). The economics of immense risk, urgent action and radical change: Towards new approaches to the economics of climate change. *Journal of Economic Methodology, 29*(3), 181–216.

Triffin, R. (1959, October 28). Statement to the Joint Economic Committee of the 87th US Congress. Reprinted 1960. Gold and the dollar crisis. New Haven: Yale University Press.

Varga, J., Roeger, W., & in 't Veld, J. (2022). E-QUEST: A multisector dynamic general equilibrium model with energy and a model-based assessment to reach the EU climate targets. *Economic Modelling, 114*(C). Elsevier.

World Bank. (2022). *Achieving climate and development goals: The financing question*, document for the October 14, Development Committee Meeting, Washington, DC. https://www.devcommittee.org/sites/dc/files/download/Documents/2022-10/Final%20Achieving%20Climate%20DC2022-0006.pdf

Part II
Financing Green Economy

The Role of European Investment Bank (EIB) and National and Regional Development Banks in the Green Transformation

Stephany Griffith-Jones and Marco Carreras

Introduction

This chapter examines the role that development banks in general, and the European Investment Bank (EIB) in particular, can play in the green transformation. It begins by discussing their renewed importance worldwide in the context of the global financial crisis of 2008/2009 and the important countercyclical role they played during COVID (2020–2022). It then proceeds to explore the long-term role they can play in the green transformation in the EU, and at the global level, in the four axes of (regional) environmental and climate change policies advanced in the introduction of this volume, i.e., regional redistribution mechanisms (including regional banks and funds from third parties), regulations, rights, and cooperation.

In recent decades, we have witnessed a real renaissance of development banks. They were much maligned in the period of the so-called Washington consensus, when private financial market efficiency was taken to the extreme, to the effect that, almost by definition, public development banks had almost no role to play, as private banks and private financial markets were perceived to do best on their own. Nowadays, the acknowledgment of the importance of development banks has been reborn.

This started with the so-called global financial crisis of 2008/2009, when public development banks at all levels, multilateral, regional, and national, significantly stepped up their lending. The World Bank in its survey (De Luna-Martinez et al., 2017) showed how National Development Banks, which quantitatively represent the most important category within the total of development banks, increased their

S. Griffith-Jones (✉)
Central Bank of Chile, Santiago, Chile
e-mail: sgj2108@columbia.edu

M. Carreras
Institute of Development Studies, Sussex University, Brighton, UK

© The Author(s) 2024
A. Ribeiro Hoffmann et al. (eds.), *Climate Change in Regional Perspective*,
United Nations University Series on Regionalism 27,
https://doi.org/10.1007/978-3-031-49329-4_6

lending by 36% in 2 years, between 2007 and 2009. Similarly, the World Bank and regional development banks also significantly increased their lending during those years. At that time, countries were desperate for funding, as in times of uncertainty, private banks and other financial institutions do not lend or lend much less, as they tend to be pro-cyclical.

The global financial crisis was followed by the Eurozone debt crisis, when again development banks stepped up and helped compensate for declines or slower growth in private lending. At that time, there was increase in the role of the EIB, in the context of the Juncker Plan (Griffith-Jones & Carreras, 2021). After the Eurozone debt zone crisis, almost all European countries created national development banks and increased the size of existing ones. The same occurred during and in the aftermath of COVID. This major countercyclical function of responding in times of uncertainty or crisis by increasing lending has made development banks greatly valued.

A further impulse to development banks came from Asia, through the creation of China-led but more broadly Asian-led, Asia Infrastructure Investment Bank (AIIB), which also has important European membership, as well as from the BRICS New Development Bank (NDB). In addition, India, which had previously closed down many development banks, is now creating one for infrastructure. This trend has also been observed in other regions. For example, in Africa, Nigeria and Ghana recently created new development banks.

These public development banks are already very important actors, especially if we add them all up, multilateral, regional, and national ones. It has been estimated that their assets amount to over US $ 23 trillion worldwide and over US $ 2.3 trillion of annual lending, which represents 10% of total global investment. Thus, development banks are already important actors. It is in this broader context that this chapter proposes to discuss the role of the European Investment Bank in the green transformation, within the EU and in the world more broadly.

Before doing so, it is important to mention the significant role the EIB played during COVID and particularly in vaccine development. For example, the EIB partially funded the initial BioNTech Pfizer vaccine, which has played such an important role in the fight against COVID in Europe and worldwide (Griffith-Jones & Carreras, 2021). This important and effective vaccine might not have come to light or been developed so quickly if the EIB had not made an initial crucial loan to BioNTech. Thus, this initial loan by the EIB had a very strong element of global public good, not only in the European context but also in the international context as well.

EIB and Green Transformation

The role of the EIB in the green transformation must be seen in the broad context of the European Union's willingness to take the lead, at a global level, to become carbon neutral by 2050, which has now been followed by other major countries including the United States. It is important to stress that this green transition implies major structural transformations across sectors in EU economies and others worldwide, including electromobility and renewable energy like green hydrogen and the corresponding infrastructure to support this (Mazzucato & Mikheeva, 2020). This implies not just new projects but also important research and development. This can be best conceived as what Mariana Mazzucato (2011) calls missions, which are cross-sectorial and involve different activities; an additional challenge is that this green transition must be just, so as to protect the poorest and the most vulnerable.

Initially, the European Commission estimated that an increased investment of at least 260 billion euros a year by 2030 was required to reach this 2050 target of carbon neutrality. But now, given the more ambitious target of reducing carbon emissions by 55% in 2030, even larger amounts are required. The European Investment Bank is central to the European Green Deal and it has committed that, by 2025, 50% of its lending will go to climate change-related activities, both mitigation and adaptation (Griffith-Jones & Carreras, 2021). Furthermore, the EIB has committed that by 2030 it will catalyze 1 trillion euros for investment in these areas.

In fact, all the operations of the EIB are under European Green Deal priorities, and the bank has stopped funding fossil fuels. The EIB has developed and used pioneering instruments like the shadow price of carbon to evaluate projects (see Stern and Stiglitz (2021), as well as the chapter by Stephan Schulmeister in this book) over a decade ago, being a true pioneer in that respect, which is interesting from an international perspective, and also of course for Latin America (Griffith-Jones & Leistner, 2018). Already in 2022, the shadow price of carbon used by the EIB reached 80 euros per ton of carbon, and it is planned to increase to 800 euros per ton of carbon by 2050, increasing gradually before. It would be valuable to discuss whether this instrument could be relevant for regional and national development banks in Latin America. In addition, there is an EU green taxonomy, which has been quite positive, even if not perfect. EIB lending has to be aligned with this EU green taxonomy. And again, this kind of instrument could be very positive to Latin American regional and national development banks. However, there remains an important challenge, which is how to make the financial intermediaries, through which the European Investment Bank channels an important part of its lending, align with the Paris criteria and with the EU green taxonomy.

The EIB has also designed an interesting climate roadmap to accelerate the green transition, to make it just, accountable, and Paris-aligned. This means not just lending to green activities but also increasing investment in innovative green technologies (Griffith-Jones and Carreras, op cit). The EIB is active, for example, in funding and promoting research on hydrogen, including green hydrogen produced with renewable energy. This has an important international dimension, including for Latin America, as Chile, for example, is very active, both through its development bank, CORFO, and via the private sector, in promoting and investing in green hydrogen, both for the domestic market but also for export, including to the European Union (Carreras et al., 2022; Griffith-Jones et al., forthcoming).

It should be stressed that EIB also has an important international dimension because 10% of its lending go to emerging and developing countries, including Latin America, and since 2020 35% of EIB lending to emerging and developing countries goes to climate finance; this proportion will be increased. Therefore, the EIB is also playing an important role in the transition to the green economy worldwide.

Furthermore, if we look at the sectors the EIB lends to, we can note that its lending is progressively becoming more aligned with the green transition. For example, the EIB has committed that, within the EU, it will not lend to any further expansion of airports unless they have a very clear green element (Climate Bank Roadmap, 2021–2025). In fact, they will only make loans for activities that mean greening airports.

On the other hand, green NGOs contest that, in certain regions and countries, the EIB still supports road building if they meet certain tests (Counter Balance (2020), Transport and Environment (2020). The EIB does not seem to value enough alternatives like trains, which are powered mainly or completely by electricity, being less carbon-intensive than road transport. The EIB argues that road building is legitimate because it supports electric cars, which means that road transport will become more low-carbon. However, there is a lot of uncertainty about the speed of developing electric cars, which leads to some controversy on this. The role that green NGOs are playing in these debates can be very constructive, because they put pressure on public development banks and on the EIB in particular to accelerate the share of their investment in genuinely low-carbon investment.

There is also a need to diversify instruments beyond credit and guarantees, such as equity. Being more involved in equity in companies allows institutions like the EIB to have greener directionality and to have more of what we call traditionally industrial policy capacity, which is very important to the green transition. Secondly, the greater use of instruments of direct equity or quasi-equity means that the EIB does not just share the risks with the private sector but also shares the potential upside, which is positive (for more details see Griffith-Jones and Carreras, op cit). One example of quasi-equity instrument used by the EIB is called venture debt. If the company does well, the EIB has the right to transform the debt into equity and therefore receive part of the profits.

An important final point is that institutions like the EIB, as well as other development banks, need not only to provide better, greener, more inclusive loans, but they need to do significantly more, especially in the area of the low-carbon transition. The scale of investment needed is so large that there is a requirement to have larger development banks. In many countries in Latin America, like Chile, the countries have good development banks, but they are very small (Griffith-Jones et al., forthcoming). Countries like Brazil, which had very important and large development banks, like BNDES especially, recently reduced its scale, when there is a need to increase its scale (Carreras, forthcoming). It is, therefore, important and cheap for governments to increase the paid in capital of development banks (Griffith-Jones et al., 2022). And in the European case, it is also important for the European

Commission to provide guarantees, which imply that the EIB can take more economic risk and fund these new technologies and their dissemination.

This is a very important lesson internationally as well. It is paramount that the capital of the World Bank is increased rapidly, as well as the capital of the African Development Bank and Inter-American Development Bank, because these countries have limited fiscal space, and they need additional international support, including for the low-carbon transition. However, the international community has been relatively inactive and somewhat silent. It has expanded liquidity for the creation of Special Drawing Rights in the wake of the COVID crisis, but not so much for funding expanded development finance.

We would like to conclude by emphasizing that there are potentially valuable lessons internationally and for Latin America in particular, from some of the pioneering aspects of the European Investment Bank, of course duly adapted to the circumstances of Latin America. These include the introduction of carbon shadow carbon pricing, green taxonomy, climate roadmap, and new instruments such as venture debt. These measures discussed above are interesting in providing experiences that can be drawn on also by Latin American and other countries.

References

Carreras, M. (forthcoming). *The Brazilian system of innovation and BNDES.* Inter-American Development Bank.

Carreras, M., Griffith-Jones, S., Ocampo, J. A., Xu, J., & Henow, A. (2022). *Implementing innovation policies: Capabilities of national development banks for innovation financing.* Inter-American Development Bank. https://doi.org/10.18235/0004390

Counter Balance. (2020). *The EIB as EU Climate Bank: Only halfway there.* Counter Balance. Brussels, Belgium. https://counter-balance.org/publications/eib-aseu-climate-bank-only-halfway-there#:~:text=Only%20halfway%20there,-%23%20EIB%20%23%20EU%20Public&text=In%20November%202019%2C%20the%20transformation,and%20environmental%20investments%20until%202030

De Luna-Martinez, J., Vicente, C. L., Arshad, A. B., Tatucu, R., & Song, J. (2017). *Survey of national development banks.* World Bank Group. http://documents.worldbank.org/curated/en/977821525438071799/2017-Survey-of-National-development

Griffith-Jones, S., & Carreras, M. (2021). *The role of the EIB in the green transformation. Policy study.* The Foundation for European Progressive Studies (FEPS) and the Initiative for Policy Dialogue (IPD), Columbia University.

Griffith-Jones, S., & Leistner, S. (2018). *Mobilising capital for sustainable infrastructure: The cases of AIIB and NDB* (No. 18/2018). Discussion Paper.

Griffith-Jones, S., Spiegel, S., Xu, J., Carreras, M., & Naqvi, N. (2022). Matching risks with instruments in development banks. *Review of Political Economy, 34*(2), 197–223.

Griffith-Jones, S., Kerrigan, G., & Petersen, J. (forthcoming). *The Chilean National System of Innovation (NSI) and the role of CORFO in innovation.* Inter-American Development Bank.

Mazzucato, M. (2011). The entrepreneurial state. *Soundings, 49*(49), 131–142.

Mazzucato, M., & Mikheeva, O. (2020). *The EIB and the new EU missions framework.* UCL Institute for Innovation and Public Purpose, IIPP Policy Report (IIPP WP 2020-17). https://www.ucl.ac.uk/bartlett/public-purpose/wp2020-17

Stern, N., & Stiglitz, J. E. (2021). *The social cost of carbon, risk, distribution, market failures: An alternative approach* (Vol. 15). National Bureau of Economic Research.

Transport & Environment. (2020). Europe's Climate Bank Six steps to transform the EIB into the world's leading bank for zero-emission transport, December 16, 2020 https://www.transporten-vironment.org/sites/te/files/publications/202012_TE_EIB_transport_lending_policy_final.pdf

Fixing Rising Price Paths for Fossil Energy: Basis of a "Green Growth" without Rebound Effects

Stephan Schulmeister

Introduction

At present, the key policy challenges are twofold, first, preventing a climate catastrophe and, second, overcoming the social and economic crisis. The first challenge calls for a reduction of (net) carbon emissions to zero as fast as possible. Reaching this target necessitates a comprehensive renovation of the capital stock[1]:

*I thank Karl Aiginger, Kurt Bayer, Michael Goldberg, Robert Guttmann, Andrea Herbst, Gustav Horn, Jürgen Janger, Jakob Kapeller, Claudia Kettner-Marx, Daniela Kletzan-Slamanig, Angela Köppl, Helga Kromp-Kolb, Timm Leinker, Ina Meyer, Manfred Mühlbacher, Nidhi Nagabhatla, Walter Ötsch, Stefan Schleicher, Mario Sedlak, Franz Sinabell, Eva Sokoll, Karl Steininger, Gunther Tichy, and Achim Truger for valuable comments.

[1] Many studies deal with pathways towards a zero-carbon economy. See the publications of the Commission on the European Green Deal and on the intermediate target of reducing CO_2 emissions until 2030 by 55% ("Fit for 55"), in particular on the investments necessary to achieve this target (European Commission, 2022). Wildauer, Leitch, and Kapeller (2020) consider a higher volume of investments necessary to reach climate neutrality than the European Commission. A much more optimistic scenario is sketched in McKinsey & Company, 2020. For a comprehensive treatment of the climate crisis in the context of environmental sustainability in general, see European Environment Agency, 2019. A roadmap for the global energy sector is provided by the International Energy Agency (IEA, 2021). For Germany, pathways towards a climate-neutral economy are investigated in Prognos et al. (2020) and in the Ariadne Report (2021). All these studies do not quantify the impact of the different "transition investments" on economic growth and, hence, do not deal explicitly with the related rebound effects.

S. Schulmeister (✉)
Independent reseacher, Vienna, Austria
e-mail: stephan.schulmeister@wifo-pens.at

- Transformation of residential and commercial buildings into little power stations through the combination of better isolation, photovoltaics, heating pumps, and batteries.
- Construction of a trans-European high-speed railway net as alternative to air travel.
- Expansion of local public transport, especially in large cities, as an alternative to private car transport.
- Replacing cars and trucks with combustion engines with emission-free vehicles.
- Moving in industrial production from using fossil energy to "green" hydrogen.
- Massive expansion of power generation from renewable sources as well as of power grids and storing capacities to meet the massively rising electricity demand.

The realization of these investment programmes would raise economic growth over the transition period of roughly 30 years. In the case of Germany, GDP would grow by roughly 3 percentage points per year higher than without such a Green Deal (as sketched in the annex). Such "green growth" would enable the renewal of the capital stock as the basis of a future circular economy. Once this is achieved, economic growth could be reduced to close to zero. Over the transition period, the "green growth" would also mitigate the social and economic crisis through providing more good jobs and financial means for modernizing the welfare state.

But what about the rebound effects of "green growth"? This issue is particularly important as using exclusively renewable electric power necessitates, e.g. the production of many times more wind power stations as already exist (as sketched in the annex). Since they consist mainly of steel and cement, their production is extremely CO_2-intensive.

This example points to the following paradox: on the way to an emission-free economy, additional CO_2 emissions must be accepted, stemming from the production of those capital goods that enable an emission-free economy in the future.

Adherents of a degrowth strategy might argue that this dilemma should be solved by shrinking production and consumption in other sectors. This conclusion is drawn from the empirical evidence: "absolute decoupling" of greenhouse gas (GHG) emissions from GDP growth (i.e. declining emissions in absolute terms) has rarely been realized in the past, certainly not as large, and fast as necessary to prevent a climate catastrophe (Haberl et al., 2020).

In the case of CO_2 emissions, however, it is particularly problematic to extrapolate from past trends to the future. First, the awareness of the danger of a climate catastrophe is much more pronounced today than it was in the past. Second, fossil energy prices and, hence, emission costs have fluctuated enormously in the past and have fallen in real terms over the long run. Hence, the profitably of emission-reducing investments has remained uncertain.

In more technical terms: for any path of economic growth, there exists a path of rising fossil energy prices so that the (demand raising) income effects of overall production are overcompensated by the (demand dampening) substitution effects of rising (relative) prices of fossil commodities. In this way, one can control and restrict the rebound effects of economic growth on CO_2 emissions and, hence, can reconcile economic growth with ecological targets.

To put it concretely: if the prices of crude oil, coal, and natural gas had risen steadily faster than the general price level over recent decades, CO_2 emissions would have become progressively more expensive. This would have incentivized business and households to save fossil energy and to invest in energy efficiency as well as in renewable energy production. In this way, carbon emissions would have been steadily decoupled from economic growth.

Unfortunately, carbon pricing through taxes or emissions trading cannot incentivize carbon-reducing investments to a sufficient extent as they cannot anchor the expectation that the *effective* costs of emissions will increase *steadily*. These effective costs consist of two components, the respective world market price of oil, coal, and natural gas and the CO_2 tax or the cost of emission certificates, respectively. If people repeatedly experience that the effective emission costs decline because world market prices of fossils and/or emission prices decline, then these expectations cannot be established.[2]

In other words, in a world of widely fluctuating prices of fossil commodities as well as of emission certificates, conventional carbon pricing cannot provide *planning security* necessary for a strong and steady expansion of carbon-reducing investments. This uncertainty problem is massively exacerbated by the extremely long payback periods of those investments.

As neither carbon taxes nor emission trading schemes can sufficiently incentivize the necessary investments in a permanent reduction of carbon emissions, this paper presents an alternative approach taking the EU and its European Green Deal as an example: the EU sets a path of steadily rising prices (e.g. by 7% per year) of crude oil, coal, and natural gas by skimming off the difference between the EU target price and the respective world market price through a monthly adjusted quantity tax. In this way, the uncertainty about future costs of carbon emissions and, hence, about the profitability of avoiding, then would be eliminated.

This chapter is structured as follows: the next section deals with the contradiction between the need for planning security of "green investments" and the price instability of fossil energy and carbon emission permits, respectively. Then, the reasons are discussed why the conventional ways of CO_2 pricing cannot incentivize green investments to an extent required for a sustained carbon reduction. The next section explains the alternative approach of fixing long-term price paths for crude oil, coal, and natural gas. Then, the political feasibility of the price path model in a (partly) de-globalizing world is examined. The final section evaluates the model as a contribution to the challenge of organizing a global "collective action" for avoiding a climate catastrophe.

---

[2] This is in no way to suggest that the current forms of CO_2 pricing do not have a dampening effect on emissions. That this is indeed the case is shown by developments in countries such as Great Britain, Sweden, Denmark, and Germany, where absolute decoupling has succeeded to a noticeable extent (for the effects of CO_2 pricing to date, see Andersson, 2019, Best et al., 2020, World Bank Group, 2020). However, much greater efforts are needed to achieve a climate-neutral economy by 2050.

Oil Price Instability and Planning (In)Security of Green Investments

Investments in energy efficiency and/or in renewable energy only pay for themselves after many years (e.g. energetic refurbishment of buildings, diffusion of electric cars, etc.) or even decades (e.g. hydrogen technology in industry, a trans-European net of high-speed trains, etc.). A successful ecological transition therefore requires *maximum planning security*.

At the same time, prices of fossil commodities, in particular crude oil, fluctuate in a sequence of bull and bear markets (typical for asset prices in general). Between 1973 and 1982, e.g. crude oil prices increased tenfold, mainly due to the two "oil price shocks" in 1973 and 1979, respectively (Fig. 1). In both cases, OPEC took advantage of political turbulences in the Middle East to "retaliate" for the preceding dollar depreciations 1971/1973 and 1976/1979, respectively (Schulmeister, 2000).

Triggered by the global recession 1980/1982, oil prices fell by more than 50% between 1980 and 1985. However, oil producers were compensated by the rising value of the dollar. When the dollar started to fall again, Saudi Arabia flooded the oil market with additional supply to restore production discipline within OPEC. This strategy failed and oil prices stagnated for roughly 15 years (Fig. 1).

After the recession of 2001, oil prices started to boom again, declined between 2011 and 2016 by roughly 70% (mainly due to additional supply stemming from fracking technologies), recovered between 2016 and 2018, and then fell again and almost collapsed in early 2020 when Saudi Arabia returned to her strategy of 1986,

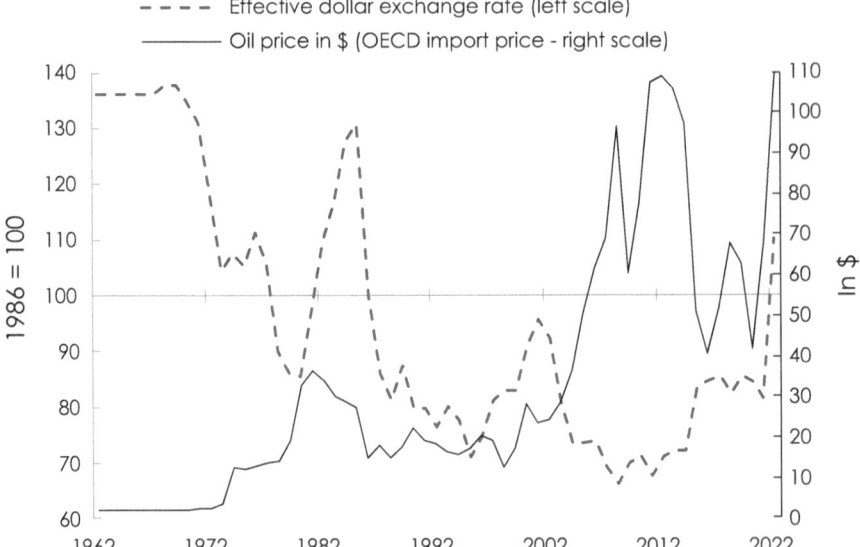

Fig. 1 Dollar exchange rate and oil price fluctuations. (Source: OECD, IMF)

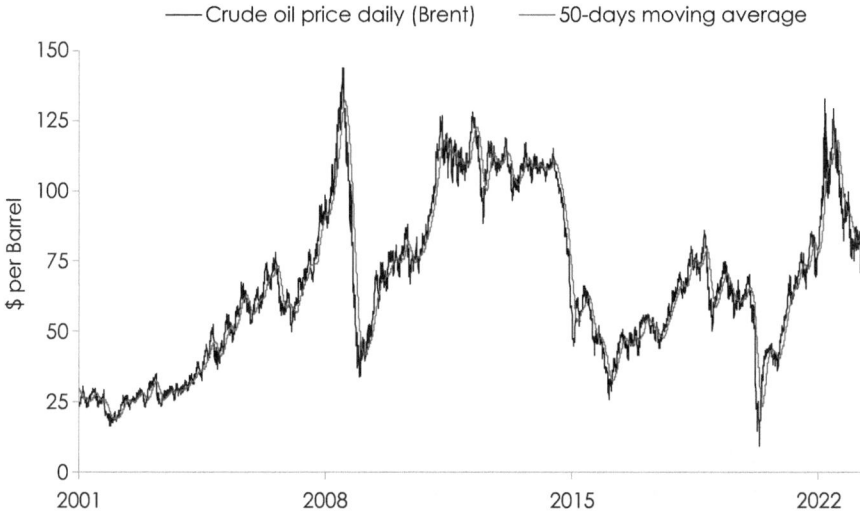

Fig. 2 Trending and speculation in the crude oil futures market. (Source: US Energy Information Administration (EIA))

i.e. flooding the market with additional oil supply (this time to "punish" Russia for not cooperating in reducing oil production). At the same time, also stock prices collapsed in reaction to the outbreak of the COVID-19 pandemic.

The unprecedented intervention by central banks in March 2020 "aborted" the bear markets and fostered a bullish sentiment in all asset markets (stocks, bonds, commodities, real estate, crypto currencies). The quadrupling of crude oil *prices* (from $20 to $80 between April 2020 and November 2021 (Fig. 2)) was additionally fostered by supply restrictions of oil-producing countries (see also section "Political Feasibility of the Price Path Model in Times of Multiple Crises"). In contrast to what was "rationally" to be expected, also EU emission prices quadrupled (Fig. 3). Fossil energy prices continued to rise after the invasion of Russia into Ukraine (in contrast to stocks, bonds, and cryptocurrencies).

As sketched above, important turning points in oil price trends were triggered by economic and political events ("fundamentals"). But why did the subsequent upward or downward trends last so long? Such an "overshooting" of asset prices can be explained as follows.

Speculative prices like those of stocks, foreign exchange, oil futures, or CO_2 emission permits fluctuate almost always around "underlying" trends (Figs. 2 and 3).[3] The phenomenon of "trending" repeats itself across different time scales ("self-similarity"). For example, there occur trends based on tick or minute data as well as trends based on daily data.

---

[3] Empirical research on the role of technical trading in asset price dynamics in general is documented in Schulmeister, 2009, as regards commodities prices, in particular oil prices, in Schulmeister, 2012.

"Technical" or "algo(rithmic)" trading aims at exploiting the trending of asset prices. In the case of *trend-following* moving average models, a trader would open a long position (buy) when the current price crosses the MA (moving average) line from below and sells when the opposite occurs (Figs. 2 and 3). By contrast, *contrarian* models try to profit from trend reversals and, hence, change open positions when a trend "loses momentum".

Technical models are applied to price data of almost any frequency. Due to the increasing use of intraday data, algo trading has become the most important driver of the rising "speed" of trading and the related boom in the volume of financial transactions.

Long-term price trends result from the following process: "mini-trends" (e.g. based on minute data) add up to one trend based on 10-minute data. Several of these trends accumulate to one trend based on hourly data and so on. Over an extended period, upward (downward) trends last longer than countermovements (mainly due to a "bullish" or "bearish" sentiment), causing the price to rise (fall) in a stepwise process. Figure 2 shows how oil price trends based on daily data accumulate to bull markets and bear markets.

The concurrence of two types of market failure in fossil energy pricing, i.e. neglect of environmental costs and "overshooting", represents fundamental causes of global warming. As a consequence, a consensus has emerged since the 1990s that CO_2 emissions should be priced, either through emission trading or through carbon taxes.[4] Unfortunately, neither instrument can ensure that the effective emission costs will steadily and permanently rise.

Carbon Pricing Through Emission Trading Systems

The EU Emission Trading System (ETS) was introduced in 2005 and covers the main CO_2 emitters from industry such as steel, paper, chemical or cement producers, as well as power generators which together account for about 45% of all CO_2 emissions in the EU.

In theory, emission trading is an optimal control instrument (see the literature mentioned in footnote 5): CO_2 emissions are limited by the volume of emissions allowances (EUA), and this cap is gradually reduced. A uniform price is formed on the permit exchanges, which ensures that the emissions take place where their benefit is greatest: A company that needs certificates buys them via the exchange from another company that has a surplus. These transactions constitute *compliance transactions*.[5]

---

[4] The general issue of carbon pricing is analysed in Edenhofer et al. (2019); Guttman (2018); Köppl, Schleicher, and Schratzenstaller (2019); OECD (2018); Sachverständigenrat (2019); and the report of the Stiglitz-Stern-Commission (2017).

[5] For an overview of the EU Emissions Trading System, see Marcu et al., 2022, European Environment Agency, 2020, and Ellerman et al., 2016. A summary of emissions trading worldwide

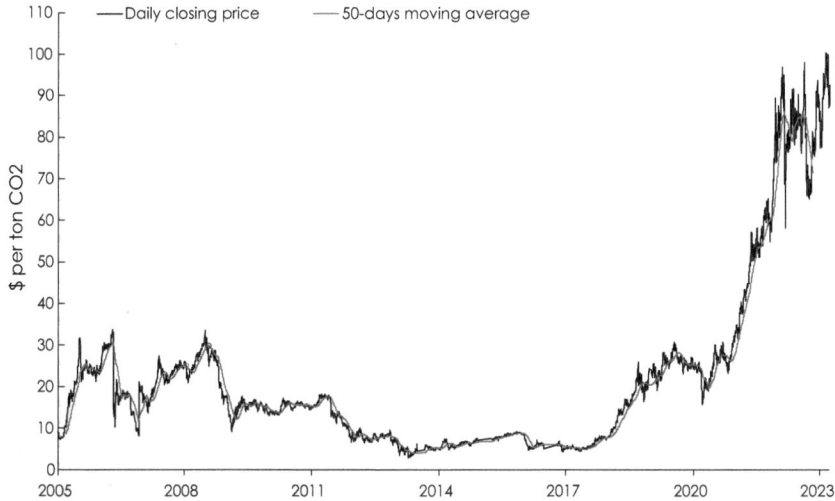

Fig. 3 Fluctuations of the futures price of EU CO_2 emission allowances. (Source: Intercontinental Exchange (ICE))

In order to incentivize sufficiently investments in carbon reduction, permit prices would have to rise steadily – at least they should not widely fluctuate. But precisely this has been the case: Since the introduction of the ETS, the price for the emission of 1 ton of CO_2 has been fluctuating between (roughly) €3 and €30. Such a low level could not provide a significant incentive to invest in reducing emissions. However, between April 2020 and December 2021, the CO_2 price rose from € 30 to roughly € 90 (Fig. 3), astonishingly in tandem with the boom of fossil energy prices (Fig. 2 – higher energy prices should have dampened the demand for emission permits). Since then, carbon prices have been fluctuating between € 70 and € 100.

This failure of emission trading to incentivize (investments in) carbon reduction to a sufficient extent has two main causes. First, the number of certificates must be fixed in advance for a longer period. This organizational necessity leads to misallocations and thus "wrong" CO_2 prices due to the fundamental uncertainty about the medium-term economic developments like a financial crisis and its effects.[6] This

---

is ICAP, 2021, and OECD, 2022. The microstructure of carbon emission markets is discussed in Kachi and Frerk, 2013, and Mizrach and Otsubo, 2014. The importance of (destabilizing) speculation in the spot and derivative markets of EU emission allowances is examined by Berta et al., 2017. Schultze (2021) provides anecdotical evidence about the rising importance of hedge funds and other financial speculators in EU emissions trading.

[6] The problem of uncertainty about the effective carbon emission costs is even bigger in the case of emission trading schemes as compared to carbon taxes as actors can know the carbon tax rate but not the future emission permit prices (Aldy and Armitage, 2020). Bayer and Aklin (2020) argue that even if carbon prices are low, an emission trading system can reduce emissions if it is a credible institution which is believed to become a more stringent in the future. They show that the EU ETS saved 3.8% of overall emissions relative to a world without carbon markets. The extent of this reduction is, however, much too low compared to what is required.

shortcoming could only be mitigated through ETS reforms (implementation of the Market Stability Reserve, reduction of the emission caps and of free permit allocation, raising the reduction targets as part of "Fit for 55", etc.).

Second, financial actors on the CO_2 permit exchanges "interpose" themselves between companies with a surplus or deficit of permits and use permit futures as vehicles for speculation. Thus, since 2010, 99% of all permit transactions have been carried out in derivatives and only 1% in genuine certificates. Already in 2012, the total CO_2 transaction volume (including derivatives) of all actors was more than 33 times higher than the companies' "compliance needs" (Berta et al., 2017) Moreover, the CO_2 price dynamics show the pattern typical for speculative prices in general: short-term trends, which are exploited by algorithmic trading, accumulate into longer-term bull or bear markets (Figs. 2 and 3).

However, most studies on the role of financial institutions in carbon trading conclude that the activities of these agents focus on hedging transactions for (polluting) nonfinancial corporations as well as on market making, i.e. providing liquidity. Speculation would play only a minor role (Ampudia et al., 2022). If this were true, then the share of outstanding open positions *between financial institutions* in the EUA futures markets should not be as big as it is, i.e. almost 50% of overall positions (see Chart B in Ampudia et al., 2022). This observation is in line with the fact that in asset markets the greatest part of liquidity is generated by "algo trading" of all kinds (between 60% and 70% of total volume). In addition, one should keep in mind that there is no precise distinction between market making (arbitrage) and speculation because the difference between the sell and the buy price reflects not only the bid/ask spread but also the speculative component due to extremely short-term price movements.

Carbon Pricing Through Emission Taxes

In all EU countries, there has long been a tax on fuels. It is equivalent to a tax on CO_2 emissions caused by fuel consumption since there prevails a fixed relationship between the quantity of fuel consumed and the related carbon emissions.[7]

In Germany, e.g. the tax on diesel is 47 cents per litre. Since the burning of one litre diesel produces 2.65 kg CO_2, the diesel tax burdens the emission of one ton of CO_2 by roughly 180 € (= 0.47/2.65 per kg). This is much more than in most planned or – like in Sweden or Switzerland – already implemented (general) carbon taxes (see Kettner and Kletzan-Slamanig 2018).[8]

---

[7] An overview of carbon taxes on CO_2 emissions from energy use in 42 countries can be found in OECD (2018). Kirchner et al. (2018) analyse the macroeconomic and distributional effects of CO_2 taxes for Austria.

[8] In fact, fuel taxes compensate also for other externalities like air pollution and noise as well as for the wear and tear of infrastructure. However, in this paper I focus on the effective costs of CO_2 emissions for households and enterprises.

Due to the volatility of crude oil prices, there occur frequently phases of markedly declining end-user prices for fuels despite (very) high-quantity taxes on fuels and – implicitly – on carbon emissions (even as high as 180 € per ton).

A concrete example illustrates the issue: between 2004 and 2008 and between 2009 and 2012, the price of crude oil rose dramatically and with it the price of fuels, heating oil, and natural gas. In Germany, e.g. the diesel price rose to € 1.50 (Figs. 2 and 4). However, the oil bull market was followed by a bear market, and the diesel price fell again to only about € 1 in 2009 as well as in 2016. Consequently, the demand for (diesel-consuming) SUVs picked up again and investments in CO_2 reductions, which were profitable at an oil price of € 70 (and more), turned into "sunk investments". In early 2020, oil and diesel prices fell once again strongly, followed by a reverse movement afterwards.

The combination of small price elasticities of both, demand and supply in oil markets, with frequent demand and supply shocks causes sharp oil price changes which are then reinforced by algo trading. Under these conditions even rising carbon tax rates cannot anchor the *expectation* of *steadily* rising paths of the *effective* costs of CO_2 emissions.

Rather the opposite could take place: the more the EU (and other countries) succeed in reducing demand for fossil energy, the more likely it is that world oil prices will fall (again, in particular as the proven reserves of oil, coal, and natural gas amount to roughly 54, 139, and 49 times global annual demand and, hence, are much larger than the global carbon budget – if a climate catastrophe is to be avoided, the reserves must not be exhausted.[9]

Fixing Long-Term Rising Paths of Fossil Energy Prices

If neither emission trading schemes nor carbon taxes can *ensure* that emitting CO_2 becomes *permanently* more expensive, how then could a rising path of effective emission costs be achieved?

The EU should set a path with steadily rising prices for oil, coal, and natural gas and skim off the difference between the EU target price and the respective world market price by means of a monthly adjusted quantity tax – instead of the end-user prices (including taxes and/or emission permit costs), the quantity tax should fluctuate. Hence, this tax can be conceived as a (implicit) carbon tax just constructed differently.

Here is a thought experiment using the example of crude oil to illustrate the working of such a price and tax regime. On January 1, 2006, the following regulation came into force in the EU: starting from the (then) current oil price (Brent) of

[9] For data on global oil, coal, and natural gas reserves, see https://ourworldindata.org/grapher/years-of-fossil-fuel-reserves-left. For a documentation of the discrepancy between countries' planned fossil fuel production and global production levels consistent with limiting warming to 1.5 °C or 2 °C, see http://productiongap.org

*Target price path: Crude oil prices in the EU rise by 5 percentage
points faster than target inflation, i.e., by 7% per year
(fictitiously from January 1, 2006).*

Fig. 4 Price incentives for CO_2-reduction – market prices versus target prices. (Source: US Energy
Information Administration (EIA), German Automobile Club (ADAC))

52.0 €, the price valid within the EU would rise along a predetermined path by 7%
per year (5 percentage points higher than target inflation).

As a result of a second bear market, the oil price fell from €95.0 to €28.3 between
March 2012 and January 2016, while the diesel price in Germany fell from €1.52 to
€0.99 (Fig. 4). However, the EU guideline price for oil would be € 102.4 in January
2016. For February 2016, the EU oil tax would thus amount to 74.1 € – 102.4 minus
28.3 – per barrel, about three times the oil bill (the figures are for illustrative purposes only; if an EU price path had been introduced, the world market price would
have been dampened further). The (final) diesel price in Germany would have risen
continuously (as Fig. 4 shows, both prices – expressed in the same currency – move
very much in tandem).

If one considers that the EU had to pay a total of € 414.5 billion in 2016 for
energy imports – almost exclusively fossil – it becomes clear that such a fossil
energy tax could yield more than € 1.000 billion in the medium and long run
(depending on the "start price"), and its returns would increase at an above-average
rate. On the one hand, the EU target price is rising, while on the other hand, the EU's
climate policy is curbing its energy imports and thus world market prices. As a
result, part of "fossil rents" would be diverted to the EU.

Technically, the implementation of such a flexible quantity tax would be simple
in the "digital age": based on the difference between the EU target price and the

world market price, the tax per unit of quantity of oil, coal, and natural gas valid in the following month is determined at the end of each month by the EU Commission and paid in the member states by producers and importers of fossil energy in the EU.

The levels from which the crude oil, coal, and natural gas price paths start as well as their annual growth rate are to be determined in a political process: the higher the priority given to incentivizing investments and consumption behaviour consistent with limiting climate change, the higher should be the initial price levels as well as their growth rate.

Since reliable expectations about the future profitability of ecological investments are the most important determinant of sustained willingness to invest, a comparatively small but permanent relative increase in the price of fossil energy could be sufficient to generate a sufficiently large volume of investment. If this turns out to be insufficient, price paths can be adjusted upwards. Since a reduction in the price of fossil energy is ruled out, the following holds: the earlier an investment is made, the greater is its profit. Such a system of pricing fossil energy would therefore initiate a long-lasting investment boom in avoiding CO_2 emissions.

Goods imported into the EU would be subject to an equivalent border carbon adjustment tax. As long as no comparable carbon taxes exist in the EU's trading partners, EU exports could be relieved from the EU fossil energy tax paid (analogous to VAT).

Technically, it would be far easier to implement just three flexible quantity taxes on oil, coal, and natural gas than managing the complex EU emissions trading scheme.

What would be the most important price and investment effects of EU target prices for fossil energies? All goods and services would become more expensive within the EU to the extent that fossil energy is used in their production – from fuels including kerosene to plastic products. Products produced with renewable energy or less energy would become relatively cheaper.

The predetermined rise of the prices of oil, coal, and natural gas will be processed in an almost Hayekian manner on the various submarkets, i.e. completely decentralized. This will eliminate the need for much regulation. If coal becomes steadily and predictably more expensive, then coal-fired power plants will be closed for cost reasons. Conversely, the increasing profitability of energy production from renewable sources will make the current system of surcharges on electricity consumers and their diversion to "green" producers obsolete.

The main impact of steadily rising fossil energy prices on CO_2 emissions will not be direct but rather indirect via the thereby induced investments. For any given capital equipment, the reaction of demand to rising prices is rather low, i.e. its short-term price elasticity. In the case of fuels, e.g. even the wide price fluctuations by 30 percentage points and more (Fig. 4) had very little impact on driving behaviour and, hence, on fuel demand. Exceptional price increases of fossil energy like in 2022 do, of course, force people to reduce their energy consumption; this effect will, however, be only temporary. By contrast, if people *know for sure* that the price of fuels will rise permanently and reliably, then a growing number will choose an electric vehicle when replacing their old car. The same reasoning holds for the investments of industry, electricity producers, or the energetic refurbishment of buildings.

Even though steadily rising fossil energy prices are not a sufficient condition to successfully fight global warming, they are a necessary condition for incentivizing all projects which will enable the transition towards a new energy system as part of a circular economy. Using part of the (enormous) returns from the fossil energy tax for long-term infrastructure projects would foster the ecological transformation (another part of tax revenues should offset the burden of energy price increases on low-income groups). These projects include the creation of a trans-European network for high-speed trains and investments in power grids as well as in hydrogen pipeline networks and in local public transportation systems.

Political Feasibility of the Price Path Model in Times of Multiple Crises

Under present conditions (high energy prices, Putin's war against Ukraine, etc.), it seems illusory to call for a steady increase in the price of oil, coal, and natural gas. The relevance of this model of carbon pricing can better be understood if one takes into consideration the systemic components of the present multidimensional crisis, in particular the relationship between global heating, the ownership of fossil energy as main polluter, and the related struggle over global income distribution between the "fossil rentiers" and the industrialized countries as the largest energy consumers. This struggle has drastically intensified in recent years:

- Either the strategy of the "fossil rentiers" to tighten their supply and keep fossil energy prices high fails (as it did in the past), then it will take the price paths to prevent the amplification of global warming through again (too) low fossil energy prices.
- Or OPEC and non-OPEC together with the transnational energy corporations succeed in forming a "quasi-cartel", then the industrialized countries need to fix rising price paths for oil, gas, and coal as a counterattack in the distribution struggle.

Looking back at the developments in recent years clarifies the issue. The Paris Agreement of 2015 took an important step to combat global warming. Slowly the "fossil rentiers" realized their business is going out of business. If the price of CO_2 were to rise steadily through taxes or emissions trading, the industrialized countries would reap the profits from the rising (gross) prices of fossil energy. It would then be difficult for the "fossil rentiers" to raise prices themselves, also because of their enormous oil, gas, and coal reserves, e.g. the global oversupply. Under these conditions, the main strategic target of "fossil rentiers" became as follows: if fossil energy must become more expensive for containing global warming, then it is up to us to raise oil, gas, and coal prices – and not industrial countries through raising carbon taxes and/or emission permit prices.

This strategy calls for a close cooperation between OPEC, non-OPEC, and energy corporations to control supply and prices of fossil energy over a transition period of several decades. As the demand for fossil energy is price-inelastic over the short and medium run, "fossil rentiers" as a "quasi-cartel" could raise prices and keep them high.

No country had and still has a greater interest in such strategy than Putin's Russia. For this is the only way she can reduce her technological backlog and pursue her world power ambitions. Therefore, the escalation of the Ukraine conflict is not only part of Putin's neo-imperial ambitions but also of his economic strategy. And this coincides with the interests of the other "fossil rentiers" and the energy corporations. The subsequent invasion of Ukraine by Russian forces increased fossil energy prices and profits of producer countries as well as of energy corporations even further. This success in turn strengthened their collusive behaviour even more. Saudi Arabia, e.g. the most important ally of the USA in the Middle East, could halve the price of oil simply by announcing an expansion of its production, but it is not interested in doing so. Reducing production volumes and profiting from price increases is much more attractive.[10]

Whether OPEC and non-OPEC countries, together with the major energy companies, can succeed in keeping fossil energy prices high through controlling supply, cannot be assessed at present. Several arguments speak against this: the level of economic development and (thus) the interests of the various producer countries vary greatly; in view of the high prices of oil, natural gas, and coal, the poorer countries will increase their production and thus exert pressure on prices. One of the most important "players", namely, Russia, could lose power in the wake of the Ukraine war and thus also in the commodities business. Moreover, Western countries could ease sanctions against Iran and Venezuela, thus increasing the global supply of fossil energy, at least in the medium term (after possible changes in the respective political regimes).

But even if the "fossil rentiers" succeed in controlling the world's supply of fossil energy, this would mean a permanent struggle over the distribution of income and power on two levels: first, on the international level between a relatively small, economically less developed group of net fossil energy exporters and the major economic blocs the USA, EU China, and Japan, and, second, within the industrial countries between the energy sector (increasingly "financialized") and the industrial and service sectors.

The most effective "counterattack" of industrial countries is to dampen demand for fossil energy, to make investments in renewable energy sources more profitable and to disincentivize investing in fossil energy. All three objectives can be achieved

---

[10] Also oil refiners and fuel distributors used Putin's war to significantly raise their profit margins: whereas the price of crude oil and diesel at the pumping station has been moving in tandem in normal times, diesel became much more expensive as compared to crude oil in the months after February 24, 2022 (Fig. 4).

by implementing the price path model as it drives a "tax wedge" between (high and rising) prices for fossil energy users and (depressed) prices for producers. If, e.g. "fossil rentiers" succeed in pushing fossil energy prices above the EU target price and keep them there, then the EU would need to shift the price path up. Otherwise, only the "fossil rentiers" would profit from rising oil, gas, and coal prices which in turn would incentivize investments into more extraction of the "toxic treasures".[11]

Fighting Global Warming, "Climate Clubs", and the Price Path Model

The most important "promotors" of the price path model would be a growing number of environmental disasters demonstrating the variety of future catastrophes due to global heating. If, e.g. during the 2020s catastrophes of various kinds become increasingly shocking and if at the same time it becomes obvious that the climate targets set for 2030 cannot be achieved, then pressures will increase to find a simple and flexible instrument for CO_2 pricing.

The price path model meets these requirements because it represents a uniform method, even though its implementation can be differentiated according to countries and economic areas (developing countries, e.g. could introduce a fossil energy price path with a lower level and/or smaller rate of growth of target prices as compared to industrial countries). If the price path model became the basic instrument of carbon pricing for a growing number of countries, it would help to overcome the biggest obstacle to limiting global warming. This obstacle is not technical but political: all important countries and regions must pull together – never before has the problem of "collective action" arisen with such force at the level of the entire planet.

In his seminal work *The Logic of Collective Action: Public Goods and the Theory of Groups*, Mancur Olson examined already in 1965 the essential problems that arise when a group wants to maintain and preserve a common good, i.e. a good from whose consumption no one can be excluded (Olson, 1965). His thoughts can be applied to the way the "world group" deals with its most important common good, the natural environment.

The focus is on the conflicts between individual and collective rationality. Thus, the larger the group and therefore the smaller the consequences of his selfish behaviour and the less conspicuous it is, the more likely an individual will not contribute anything to the preservation of the common good, i.e. act as a "free rider". In a small group, such as a farming community, "free riding" can therefore be contained in terms of a common at the local level (Ostrom, 1990; for an application to climate

---

[11] Even considering the efforts to fight global warming, OPEC expects in its forecast (OPEC, 2022, Table 2.1) that global oil demand and production will rise by 12.3% between 2022 and 2045. The share of fossil energy in world primary energy demand would only decline from 80.2 (2021) to 69.6% (2045) – a catastrophic development for the climate. However, if the problem of a global collective action is not successfully tackled, this forecast is plausible.

change see Harris, 2007). This, however, is not true as regards preserving biodiversity at a regional level or the climate at the global level. Hence, the climate crisis can be conceived as a "tragedy of the commons" on a planetary scale.

Incentives for preserving and cultivating common goods are usually provided by the state, for example, through taxes or subsidies; however, there is no "world state" that could protect the climate. Hence, at the global level, the greatest progress has been made only in diagnosing the problem, e.g. by the International Panel on Climate Change (IPCC). Policy has yet only set targets without binding and verifiable agreements on how these targets will be achieved (as in the Paris Agreement of 2015).

This problem is deepened by the fact that the conflict between individual and collective rationality also arises at the international level in the form of national self-interest on the one hand and the global commons on the other hand: if there is no consensus on the method of combating carbon emissions, each country will choose those ways which also serve its national interests. The idea that nation-states compete against each other on a global level like companies rather than cooperating with each other as partners reinforces this danger.

Felbermayr (2021) gives a realistic example. If one country (e.g. the EU) increases the relative price of fossil energy compared to renewable energy through taxes and another country (e.g. the USA) increases it to the same extent by subsidizing renewable energy, this has very different consequences for the economies of the two countries, in terms of both their international competitiveness and their internal distribution of income.

It would therefore be ideal if, as a first (major) step towards harmonizing methods to combat CO_2 emissions, the three largest emitters, China, the USA, and the EU, were to agree on common price paths for oil, coal, and natural gas and on corresponding carbon border adjustment taxes to prevent "carbon leakage" to countries with no or low CO_2 taxation (the idea of "climate clubs" stems from William Nordhaus and has been adapted to fit the WTO rules; see Tagliapietra and Wolff, 2021; Felbermayr, 2021).[12] Exports of non-member countries to the "club" would be burdened by a border adjustment tax.

The efforts of these countries to reduce carbon emissions would be strengthened significantly if they could be convinced to also introduce price paths for fossil energy – otherwise they would have to pay the carbon adjustment tax for their exports to the "club", the most important markets for exports of developing countries. In addition, also a group of emerging market economies, e.g. the Mercosur countries, could form a "climate club" to complement economic cooperation with a common form of carbon pricing which would not affect the intraregional price competitiveness (as in the case of different national carbon taxes). At the same time, the "Mercosur climate club" could deal with the "China-US-EU climate club" about a

---

[12] Harmonizing the effective carbon prices between the member countries would provide a level playing field also within the club. This would not be the case if, e.g. China burdens CO_2 to a lesser extent than the EU. In this case, China would enjoy a comparative price advantage (only imports from non-members would be treated equally).

differentiation of fossil energy price paths. Generally, the shape of the price paths should account for the different level of carbon emissions as well as of economic development: the higher the level of emissions and of GDP per capita, the higher should be the starting level and the rate of increase of the price paths. In contrast to other forms of carbon pricing, fossil energy price paths can easily be implemented, adjusted (if necessary), and controlled.

Under these conditions, global demand for fossil energy could be steadily dampened and, hence, also carbon emissions. At the same time, also supply would be dampened as the price path model drives a wedge between steadily rising fossil energy prices for consumers/users (to dampen demand) and low prices for producers (to dampen supply).

Achieving a circular economy necessitates not only a permanent carbon emission reduction but also a steadily rising share of recycling of raw materials of all kinds. This is more important as the "material consumption" in the EU amounted to 13.4 tons per person in 2020.[13] Only 30% of the waste left at the end of the production process is recycled ("output recycling rate") or 10% of the overall material consumption ("input recycling rate").

Even though the most important instruments for raising the recycling rates consist of regulations with respect to the product design (durability, reusability, reparability), economic incentives also play a role, in particular the development of the prices of raw materials as production input. If, e.g. plastic producers *know* that crude oil prices will permanently rise faster than the general price level, then investing in more recycling capacity becomes reliably profitable. This argument holds for recycling in general as the profits of the respective investments consist primarily of the saved raw material costs. As in the case of fossil energy, setting rising price paths of (recyclable) raw materials would anchor the respective expectations.[14]

Finally, a remark to those who are convinced that degrowth is "the" necessary condition for a transition towards a circular economy. For me, economic growth is by no means an intrinsic value. Economic activities should aim at providing the basis for a good life of the greatest possible number of people. At present, the biggest challenge is organizing a collective action at the global level to fight the climate crisis. The necessary renovation of the capital stock as one fundament of a future circular economy implies huge investment programmes which would contribute to economic growth and cause additional carbon emissions. This effect could and should be (over)compensated by steadily rising prices of fossil energy.

This combination of a *transitory* "green growth" and rising fossil energy price paths seems much more in line with the goal of providing the basis for a good life of the many than shrinking economic activities in other sectors of the economy (not to speak about other parts of the world like the global South). Such a degrowth strategy would call for a radical change of the economic system as regards the

---

[13] See https://ec.europa.eu/eurostat/web/products-eurostat-news/-/ddn-20210713-2.

[14] A plan for the transition towards a circular economy in the EU (though without considering the role of raw material prices) is sketched in European Commission (2020).

distribution of working time, income, wealth, and political power.[15] Given the extremely unequal distribution of economic and political power at present, striving for a radical change of both, the ecological and the social-economic system, seems to me a mission impossible. Hence, at present one should focus on fighting global warming through the combination of "green growth" and rising price paths of fossil energy. The related creation of "good" jobs might then – gradually – also mitigate the social crisis.

Concluding Remarks

This chapter proposes a new approach to pricing CO_2 emissions: setting a path of steadily rising prices of crude oil, coal, and natural gas by skimming off the difference between the target price and the respective world market price through a monthly adjusted quantity tax. In this way, the uncertainty about future price developments of crude oil, coal, and natural gas and, hence, of the effective emission costs would be eliminated. Firms and households could calculate the profitability of investments in avoiding carbon emissions. By contrast, neither carbon taxes nor emission trading schemes can provide such a planning security, indispensable for successfully combatting global warming. The price path model of efficient carbon pricing could be implemented, e.g. by the EU but could also serve as a common basis for "climate clubs", initially comprising the greatest carbon emitters, i.e. China, the USA, and the EU, potentially followed by groups of emerging market economies like the Mercosur countries.

At first glance, fixing a path of steadily rising fossil energy prices by means of economic policy might appear as falling back to a "centrally planned economy". However, if one takes into consideration the causes of global warming, the specific conditions in (derivatives) markets for fossil energy and carbon emission permits as well as the theory of externalities and public goods, then the proposal should appear worth being discussed. The global natural environment is the most valuable common good of mankind. Confronted with the threat of its destruction, the courage to escape from conventional modes of thinking should not be lacking.

To put it in the words of Albert Einstein: "You can never solve a problem on the level on which it was created".

---

[15] For a primer in degrowth economics, see Kallis et al. (2018), Schmelzer et al. (2022), and Priewe (2022).

Annex: A Back-of-the-Envelope Estimation of the Growth Effects of a Decarbonization of the German Economy

The purpose of this exercise is to gauge in an extremely rough manner how the investments necessary to achieve a carbon-free economy might impact upon economic growth. The more the ecological renovation of the capital stock would induce a significant "green growth", the more important an effective carbon pricing becomes.

As a first step, I take estimates of the additional electricity production needed for a decarbonization of the German economy in general and its industry in particular. I estimate the number of additional wind turbines which could produce the required power as well as the costs of their installation (as regards the rated power, effective electricity production, and investment costs, I use data for the already existing wind power stations in Germany). As power production costs (per KWh) are roughly the same for wind, solar, and biogas installations (Fraunhofer, 2018), this assumption simplifies the estimation of overall power plant investment costs. Based on the results of another study, I present estimates of renewable power demand and investment costs of a decarbonization of German industry.

As a second step, I estimate the volume of investments needed to replace combustion engine cars and trucks with electric vehicles, to energetically refurbish residential buildings, and to contribute to the enlargement of the European high-speed railways net.

Power Production and Installation Costs of Wind Turbines in Germany

Power production:

(https://www.wind-energie.de/themen/zahlen-und-fakten/ – retrieved September 25, 2021 – numbers are rounded)

Number: 31.100.

Total rating power: 63 GW.

Total production: 132 TWh.

Ø Rating power: 2,03 MW (=63.000/31.000) = 2.026 KW.

Ø Production: 4,24 GWh (=132.000/31.100).

Investment costs:

(Fraunhofer ISI, 2018)

Costs per KW rating power: 2.030 € (weighted average of the average costs of onshore and offshore turbines).

Ø Costs per installation: 4112 Mill. € (= 2.030 € * 2.026 KW)

Renewable Power and Investments Needed for a Climate-Neutral German Economy

In a comprehensive study, a consort of many research institutions investigates decarbonization pathways of the German economy (Ariadne-Report, 2021). As regards the power production necessary to achieve this target, different models arrive at estimates between 639 and 1.480 TWh (Ariadne-Studie, 2021, p. 19). Taking the mean value of 1.055 TWh and subtracting the actual production volume in 2020 of 251 TWh, I arrive at an estimate of roughly 800 TWh as additionally needed renewable power. The estimate of another study (632 TWh) is smaller but not completely at odds with the Ariadne study (Prognos et al., 2020, Fig. 8).

Additional power from renewable resources: 800 TWh.

Number of additional wind turbines: 188.679 (= 800.000 GWh/4,24 GWh).

Investment costs: 774 bn. € (= 188.679* 4,1 Mill. €) = 22,1% des BIP (2021: 3.500 Mrd. €)

Investments Needed for Carbon-Free Buildings

Single-family homes (40% of population).

Number: 16 Mill.

Estimated average costs of a complete energetic renovation, i.e. combining better isolation, photovoltaics, heat pumps, and batteries: 60.000 €.

Total investment costs: 960 bn. € (= 60.000 * 16.000.000).

Apartment buildings (including houses with only few flats – 60% of population).

Here, I operate with an extremely rough estimate since apartment buildings differ very much from one another as regards size, quality of isolation, heating system, etc. Considering that a complete energetic refurbishment of apartment buildings is more expensive as compared to single-family homes (per m² living space) and that roughly 60% of the population live in apartment houses, I use as estimate of overall investment cost 1.500 bn. €.

Investment cost of renovating all residential buildings: ~2.460 bn. €.

Commercial buildings.

As their floor space in Germany amounts to 10% of the overall floor space of residential buildings, I take 10% of the renovation costs of residential buildings as estimate for commercial buildings, i.e. 246 bn. €.

Estimate of renovation costs of all buildings: 2.706 bn. € or 77,3% of GDP

Investments Needed for Carbon-Free Road Transport

If one assumes that an electric car costs on average 20.000 € more than a combustion engine car and that the stock of passenger cars falls from 48 mill. to 25 mill. Between 2020 and 2050, then additional investment costs can be estimated at 500 bn. € or 14.3% of GDP.

For electric and hydrogen trucks, additional costs can be estimated at 50.000 € per truck. If the number of trucks declines until 2050 from 3.5 mill. to 2 mill. Due to shifting goods transport to railways, then overall additional investment costs can be estimated at 100 bn. € or 2,9% of GDP.

Investments Needed for the Enlargement of a Trans-European High-Speed Railway Network

As part of the construction of a European Green Deal, the high-speed railway network should be accelerated. If additional 30.000 km would be constructed (at present: 10.000 km), then investment costs would amount to 600 bn. € in the EU or 4.3% of GDP of the EU (according to the International Union of Railways, construction cost per km vary in Europe between 12 and 30 mill. €; assuming 20 mill. €, one arrives at overall cost of 30.000 * 20 = 600 bn. €).

If Germany contributes an equivalent share to the European railways network, then the respective investments would amount to 4.3% of its GDP.

Overall Costs of Investments in the Transition Towards a Climate-Neutral Economy in Germany

The above back-of-the-envelope estimates sum up to 120.9% of German present GDP (2021). If all these investments were continuously carried out until 2050, they would "ceteris paribus" raise economic growth by 2.7 percentage points per year. The actual growth effect of complete decarbonization of the German economy would be higher since the above estimation exercise did not account for investments in energy storage (beyond batteries in buildings); in energy distribution through additional power grids and hydrogen pipelines; in the production of biofuels, in particular for aircrafts (and the related retrofits); in improvement of local public transportation (in particular in metropolitan areas); in reducing emissions in agriculture (biogas plants); and in carbon capture and storage. A complete decarbonization of the German economy would therefore raise economic growth over roughly three decades by 3.0 to 3.5 percentage points per year.

There are two reasons why the potential growth effects of a transition towards a climate-neutral economy were estimated in this annex. First, studies which sketch or even elaborate in detail the respective pathways assume a certain GDP growth over the transition period without analysing the feedback of the emission-reducing investments on overall growth. Prognos et al. (2020), for example, assume a growth rate of 1.3% per year until 2050 which seems inconsistent with the size of the necessary investment programmes as elaborated in their study.

Second, the results of the estimation of the growth effects of decarbonizing the economy suggest that the income effects on additional carbon emissions would be massive. Hence, emissions can only be steadily reduced through a simultaneous substitution effect of permanently and sufficiently rising prices of fossil energy (overcompensating the income effects).

References

Aldy, J. E., & Armitage, S. (2020). "The cost-effectiveness implications of carbon Price certainty", AEA papers and proceedings. *American Economic Review*, 113–118.

Ampudia, M., Bua, G., Kapp, D., & Salakhova, D. (2022). The role of speculation during the recent increase in EU emission allowances prices. *ECB Economic Bulletin, 3*.

Andersson, J. J. (2019). Carbon taxes and CO_2 emissions: Sweden as a case study. *American Economic Journal: Economic Policy, 11*(4).

Ariadne Report. (2021). *Deutschland auf dem Weg zur Klimaneutralität 2045 – Szenarien und Pfade im Modellvergleich*.

Bayer, P., & Aklin, M. (2020). The European Union emissions trading system reduced CO2 emissions despite low prices. *Proceedings of the National Academy of Sciences of the United States (PNAS), 117*(16), 8804–8812.

Berta, N., Gautherat, E., & Gun, O. (2017). Transactions in the European carbon market: A bubble of compliance in a whirlpool of speculation. *Cambridge Journal of Economics, 41*, 575–593.

Best, R., Burke, P. J., & Jotzo, F. (2020). Carbon pricing efficacy: Cross-country evidence. *Environmental and Resource Economics, 77*, 69–94.

Edenhofer, O., Flachsland, C., Kalkuhl, M., Knopf, B., & Pahle, M. (2019). Optionen für eine CO_2-Preisreform, MCC-PIK-Expertise für den Sachverständigenrat zur Begutachtung der gesamtwirtschaftlichen Entwicklung. In *Mercator research institute on global commons and climate Change (MCC)*. Potsdam-Institut für Klimaforschung (PIK).

Ellermann, A. D., Marcantonini, C., & Zaklan, A. (2016). The European Union emissions trading system: Ten years and counting. *Review of Environmental Economics and Policy, 10*(1), 89–107.

European Commission (EC). (2020). *Circular economy action plan*.

European Commission (EC). (2022). Towards a green, digital and resilient economy: Our European growth model, Brussels, 2.3.2022 COM. 83 final.

European Environment Agency (EEA). (2019). *The European environment – State and outlook 2020*.

European Environment Agency (EEA). (2020). The EU emissions trading system in 2019: Trends and projections, Copenhagen.

Felbermayr, G. (2021). Steuerliche Aspekte der Klimapolitik. *Wirtschaftsdienst, 6*, 428–431.

Fraunhofer ISE. (2018). Stromgestehungskosten erneuerbare Energien.

Guttmann, R. (2018). *Eco-capitalism – Carbon money, climate finance and sustainable development*. Palgrave Macmillan.

Haberl, H., Wiedenhofer, D., Virag, D., Kalt, G., Plank, B., Brockway, P., Fishman, T., Hausknost, D., Krausmann, F., Leon-Gruchalski, B., Mayer, A., Pichler, M., Schaffartzik, A., Sousa, T., Streeck, J., & Creutzig, F. (2020). A systematic review of the evidence on decoupling of GDP, resource use and GHG emissions, part II: Synthesizing the insights. *Environmental Research Letters, 15*, 065003.

Harris, P. (2007). Collective action on climate change: The logic of regime failure. *Natural Resources Journal, 47*, 195–224.

ICAP (international carbon action partnership). (2021). Emissions trading worldwide: Status Report 2021.

International Energy Agenc (IEA). (2021). Net Zero by 2050.

Kachi, A., & Frerk, M. (2013). *Carbon market oversight primer*. International Carbon Action Partnership (ICAP).

Kallis, G., Kostakis, V., Lange, S., Muraca, B., Paulson, S., & Schmelzer, M. (2018). Research on degrowth. *Annual Review of Environment and Resources*, 291–316.

Kettner, C., & Kletzan-Slamanig, D. (2018). Carbon taxation in EU Member States: Evidence from the transport sector. In S. E. Weishaar, L. Kreiser, J. E. Milne, H. Ashiabor, & M. Mehling (Eds.), *The green market transition, chapter 2* (pp. 17–29). Edward Elgar Publishing.

Kirchner, Mathias, Sommer, Mark, Kettner-Marx, Claudia, Kletzan-Slamanig, Daniela, Köberl, Katharina, Kratena, Kurt CO_2 tax scenarios for Austria – impacts on household income groups, CO_2 emissions, and the economy, WIFO Working Paper 558, 2018.

Köppl, A., Schleicher, S., & Schratzenstaller, M. (2019). *Policy Brief: Fragen und Fakten zur Bepreisung von Treibhausgasemissionen*. WIFO.

Marcu, A., Hernandez, J. F. L., Alberola, E., Faure, A., Qin, B., O'Neill, M., Schleicher, S., Caneill, J.-Y., Bonfiglio, E., & Vollmer, A. (2022). *State of the EU ETS Report, European Roundtable on Climate Change and Sustainable Transition (ERCTS), BloombergNEF*. Wegener Center, ecoact (eds.).

McKinsey&Company. (2020). Net-Zero Europe, December.

Mizrach, B., & Otsubo, Y. (2014). The market microstructure of the European climate exchange. *Journal of Banking & Finance, 39*, 107–116.

OECD. (2018). *Effective carbon rates 2018 – pricing carbon emissions through taxes and emissions trading*.

OECD. (2022). *Pricing greenhouse gas emissions: Turning climate targets into Climat actions*.

Olson, M. (1965). *The logic of collective action: Public goods and the theory of groups*. Harvard University Press.

OPEC. (2022). World Oil Outlook 2045.

Ostrom, E. (1990). *Governing the commons: The evolution of institutions for collective action*. Cambridge University Press.

Priewe. (2022). Growth in the ecological transition: Green, zero or de-growth? *European Journal of Economics and Economic Policies: Intervention, 19*(1), 19–40.

Prognos, Öko-Institut, Wuppertal-Institut. (2020). Towards a Climate-Neutral Germany. Executive Summary conducted for Agora Energiewende, Agora Verkehrswende and Stiftung Klimaneutralität.

Prognos, Öko-Institut, Wuppertal-Institut, Klimaneutrales Deutschland. (2020). Studie im Auftrag von Agora Energiewende, Agora Verkehrswende und Stiftung Klimaneutralität

Schmelzer, M., Vetter, A., & Vansintjan, A. (2022). *The future is degrowth*. Verso.

Schulmeister, S. (2000). Globalization without global money: The double role of the dollar as national currency and as world currency. *Journal of Post Keynesian Economics, 22*(3), 365–395.

Schulmeister, S. (2009). Profitability of technical stock trading: Has it moved from daily to intraday data? *Review of Financial Economics, 18*(4), 190–201.

Schulmeister, S. (2012). *Technical trading and commodity Price fluctuations*. WIFO Study.

Schultz, Stefan. (2021). Wie Hedgefonds den Kohleausstieg befeuern, DER SPIEGEL, 29. https://www.spiegel.de/wirtschaft/service/emissionshandel-wie-hedgefonds-den-kohleausstieg-beschleunigen-a-44bf3116-4557-4f05-b1c3-f7a4944f7be3

Stiglitz-Stern-Commission. (2017). Report of the high-level commission on carbon prices, international bank for reconstruction and development and international development association/ The World Bank, May 29.

Tagliapietra, S., & Wolff, G. B. (2021). Form a climate club: United States, European Union and China. *Nature, 59*(1), 526–528.
Wildauer, R., Leitch, S., & Kapeller, J. (2020). *How to boost the European Green Deal's scale and ambition, Foundation for European progressive studies*. Arbeiterkammer, Renner-Institut.
World Bank Group. (2020). *State and trends of carbon pricing 2020*.

Part III
New Green Solutions to Climate Change

Challenging the Status Quo: A Critical Analysis of the Common Agricultural Policy's Shift Toward Sustainability

Yannis E. Doukas, Ioannis Vardopoulos, and Pavlos Petides

Charting the Course for a Greener Common Agricultural Policy

The Common Agricultural Policy (CAP) in the European Union (EU) has various costs and benefits associated with it, and different EU Member States have divergent agendas and expectations regarding it. The CAP involves multiple actors who participate in its formation, including bureaucrats, sectoral interests, governmental agendas, and pressure organizations interested in agriculture. This shared and binding policy significantly impacts how benefits are divided among these actors.

In addition to economic benefits, the CAP also integrates social and environmental aspects to promote a resilient and sustainable-oriented agricultural structure throughout the EU. This policy encourages favorable environmental conditions that allow farmers to benefit from natural resources and maintain financial stability by producing agro-food. Agricultural income not only supports farming households and communities in rural areas but also contributes to society's overall gains from agricultural production (European Commission, 2021). In addition, agricultural

Y. E. Doukas (✉)
Department of Agricultural Development, Agri-Food and Natural Resources Management, University of Athens (UoA), Psachna, Greece
e-mail: jodoukas@pspa.uoa.gr

I. Vardopoulos
School of Environment, Geography and Applied Economics, Harokopio University of Athens (HUA), Kallithea, Greece

Department of Regional and Economic Development, Agricultural University of Athens (AUA), Amfissa, Greece

P. Petides
Research Center for Economic Policy, Governance and Development, University of Athens (UoA), Athens, Greece

© The Author(s) 2024
A. Ribeiro Hoffmann et al. (eds.), *Climate Change in Regional Perspective*, United Nations University Series on Regionalism 27, https://doi.org/10.1007/978-3-031-49329-4_8

activities are highly susceptible to climate change, but they can also help mitigate it by reducing greenhouse gas emissions and storing carbon while continuing to produce food.

The CAP policymaking process has been evolving toward the creation of a sustainable and eco-friendly framework for European agriculture, with an emphasis on promoting environmental consciousness at all levels of governance within the EU's complex and multifaceted system. This chapter delves into the theoretical underpinnings of policy change, specifically examining the concepts of neo-institutionalism and historical institutionalism, to better understand the mechanisms driving this shift. These theories emphasize the role of institutional factors in shaping policy outcomes and the importance of historical context in understanding policy development.

Dynamics of Institutional Shifts: Exploring the Mechanisms of Change

For many years, scholars have studied organizations and their interconnections. During the second half of the nineteenth century and the first part of the twentieth century, social theorists created a body of literature (Meyer, 2017; Scott, 2004). They concentrated on how institutionalization is fostered by the bureaucratic structures of organizations, as well as the organizational structure within society (Hall & Taylor, 1996). Until the 1950s, political science was primarily concerned with analyzing the structures of government and state in the United States and the United Kingdom. This earlier approach, known as "old institutionalism," focused on the formal structures and institutions that make up these political entities (Andreou, 2018). However, a behavioral movement that popularized new theories on how policies are created and changed, including behaviorism, positivism, and rational choice theory, emerged to complement it (Ostrom, 1998; Strom, 1990). This led to the limiting emphasis on institutions being dropped in favor of evaluating people instead of the institutions surrounding them (Tina Dacin et al., 2002). In 1977, John W. Meyer and Brian Rowan's significant work transformed institutionalism (Andreou, 2018), leading to a resurgence of the subject in the ensuing decade, with various fields outside the social sciences contributing to its deluge of writing.

Neo-institutionalism was coined in 1983 by March and Olsen, distinguishing it from old institutionalism (March & Olsen, 1983). Neo-institutionalism focused on comparative investigations and the independent impacts of institutions on political behavior and results (Thelen & Steinmo, 1992). It emerged as a means of expressing disagreement with the dominant at the time behaviorist theoretical currents (Kraatz & Zajac, 1996). Behaviorist perspectives overestimated the degree that institutions have a direct impact on politics, ignoring that political institutions are more than passive platforms for political conduct, but also elements that influence political behavior (Andreou, 2018). Historical institutionalists have emphasized the idea that each and every political action takes place within the context of time and that history influences decisions, actions, and occurrences in the future (Christiansen & Verdun, 2020). This perspective considers history not as a collection of specific events but as a factor that shapes policy change (Hall & Taylor, 1996). The terms "new institutionalism" and "historical institutionalism," which are characterized as progressive changes to existing institutions or new and inventive policies and their connection to reforms in policy (Bennett & Howlett, 1992), are highlighted in this study.

Historical institutionalism faced the challenge of institutional change despite its emphasis on institutional constancy (Pierson, 2000). A useful perspective for analyzing the persistence or absence of policy change may be provided by the published scholarly literature on route dependency (ibid). Path dependency refers to the fact that institutions, once established, tend to follow historically set, specific courses, making it costly to change course (Levi, 1997). Institutions "lock in" and develop along the trajectories that govern their dependency, including unintended outcomes and inefficiencies (Andreou, 2018). As a consequence, even when the existing model is inadequate, key players tend to defend it since it matches the needs of its founders, and altering policies is typically difficult due to institutions' resilience (Greener, 2002). Due to the encouragement of policy continuity by earlier decisions, public policies, and formal frameworks often turn out to be difficult to modify (Pierson, 2000).

As rational actors incorporate into the institutional environment, its effects grow, and systemic factors increasingly constrain and define the strategic options available to them (Beckert, 1999). Therefore, institutional change may occur in a specific environment whose breadth and attributes were influenced by earlier political and institutional decisions (Clemens & Cook, 1999). The idea of "punctuated equilibrium" was extensively studied in historical institutionalism examinations of institutional evolution (Thelen, 1999). Institutions tend to be in a condition of equilibrium throughout the majority of their history, operating based on the decisions made at the time of their founding or the most recent punctuation point. The "punctuated equilibrium" emphasizes how crucial the institutional atmosphere is in influencing policy dynamics and the success of future reforms (Romanelli & Tushman, 1994). Long periods of institutional stability and adherence to historical patterns can lead to critical junctures, where pivotal decisions are made that can greatly impact the trajectory of a system or policy (Collier & Collier, 1991). When opportunities for significant institutional transformation are both evident and practical, a critical juncture is characterized to be a brief span of time whereby uncertainties regarding the prospects of an institution create the grounds for policymakers to place the institution on an alternative path of growth and development (Hall & Taylor, 1996). However, a critical juncture point does not always happen at a moment when its effects may be seen in retrospect (Capoccia, 2015). Thelen and Steinmo contend that the critical juncture actually happens much sooner in the process, prior to its impacts becoming apparent (Thelen & Steinmo, 1992). "Short periods" term relates to how briefly institutions may alter their direction before reverting to their prior

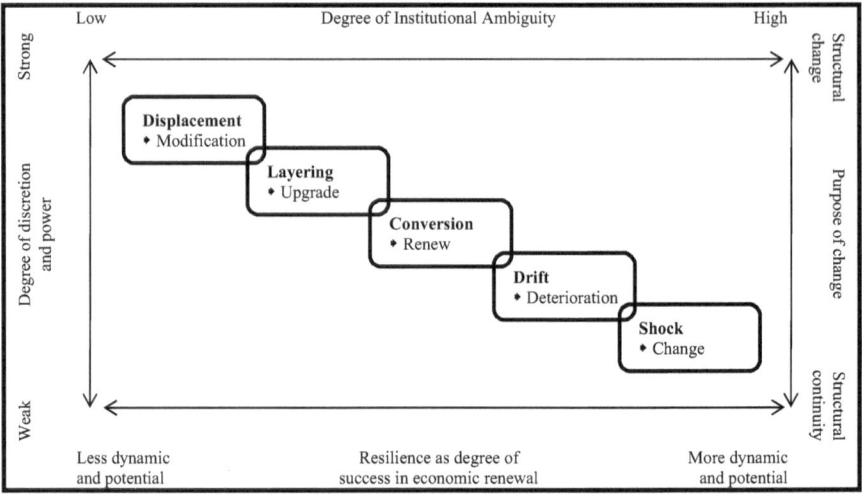

Fig. 1 Institutional change model. (Adapted from Hu and Yang (2019))

dependencies. Actors can select how an institution should evolve over time and what new policies to put in place to deal with new issues (Di Bartolomeo et al., 2021).

In their systematic approach to institutional transformation, Streeck and Thelen (2005) added additional concepts to the conceptual framework of the historical institution. According to their research, policies that establish norms, assign rights and duties to players that are backed by norms, and enable third-party enforcement can also be thought of as institutions (ibid). The theory of institutional transformation can therefore be a theory of policy change (ibid). Thus, policies can constitute rules for players other than policymakers themselves, and if required, they will be enforced by agents acting in the interests of society (ibid). This gives policies the status of legitimate institutions (ibid). According to Streeck and Thelen (2005), there are five institutional change categories: displacement, layering, conversion, drift, and shock (ibid) (see also Fig. 1).

Displacement refers to the gradual modification of existing regulatory structures (ibid). This is the most basic form of institutional change and occurs when preexisting arrangements are contested or ignored in favor of new institutions and related behavioral logic (ibid). The second category, layering, involves actively supporting changes made to an established group of institutions by adding new regulations and/ or institutions alongside or over the older ones (ibid). Andreou (2018) suggests that in highly partisan states, layering is a prevalent practice where new frameworks and/ or laws are created to "control" party supporters without directly affecting state institutions. This creates tension between institutions and policies that can lead to conflicts and institutional change. Moreover, recent studies have shown that the COVID-19 pandemic has created a shock to institutional structures, resulting in significant policy changes – and more – across the globe (Alaimo et al., 2022; Bol et al., 2021; Chantzaras & Yfantopoulos, 2022; Trebesch et al., 2020). This sudden and unexpected shock has forced governments to implement new policies and regulations in order to address the health and economic consequences of the pandemic (Katsikas, 2022).

Streeck and Thelen (2005) highlighted the possibility of institutional drift, gradual deterioration, or atrophy of institutions due to the failure to adapt to shifting political and economic conditions. This drift may be caused by gaps in regulations, and political maturity is essential to implementing necessary changes (Streeck & Thelen, 2005). The neglect may or may not be intentional, but institutions may be diverted to new goals, functions, or purposes due to fresh environmental issues, changes in the balance of power, or political struggles (Polishchuk, 2012). As a result, unexpected outcomes are to be expected, and change requires compromise, which may take time. Additionally, in contrast to the other four shift cycles, Streeck and Thelen (2005) identified exhaustion as a process that causes failure and can happen when an institution's normal activities degrade its external environment and available resources (ibid). It usually happens gradually rather than suddenly (ibid). When an institution experiences exhaustion, activities within the organization degrade its functioning, contrary to drift, where the formal integrity of an organization is retained, despite becoming progressively dysfunctional (ibid).

Recent studies have emphasized the importance of institutional resilience, which refers to an institution's ability to adapt to changing circumstances and maintain its functions and goals (Barasa et al., 2018). Institutional resilience can be achieved through proactive strategies such as scenario planning and regular assessments of the institution's performance (Lengnick-Hall et al., 2011). Furthermore, institutional resilience can be strengthened through the establishment of flexible structures and decision-making processes that allow for quick adaptation to changes (Carmichael, 2015). Finally, studies have shown that successful institutional adaptation requires the involvement of multiple stakeholders, including policymakers, private sector actors, and civil society organizations (Wamsler, 2017). The engagement of diverse perspectives can enhance the institution's capacity to identify and respond to emerging challenges (Holley, 2009).

Extensive research in public policy has thoroughly examined the significance of concepts and knowledge in systemic change. According to this approach, a significant part of the political debate is a stage of advancement in society expressed through public policy. With the previous policy having the most significant cognitive influence, and the current policy (in time t1) reacting to the effects of earlier initiatives, public policy functions as an educational endeavor, as a means of learning. Hall (1993) refers to this mechanism as "the purposeful endeavor to adjust the goals or tactics of public policy in order to conform with old knowledge and new facts" which he defines as social learning. The majority of those involved in this learning process are professionals in public policy who advise or serve the public sector from high-ranking positions at the intellectual subcultures of society and bureaucracy nexus. Three stages make up the process of changing public policy: the broad objectives that guide policy in a given area, the methods or tools employed to accomplish those objectives, and the real costs associated with those methods and instruments. Each stage is composed of an equal number of variables. Historical institutionalists contend that the creation of institutions and policies frequently leads to conflict between groups with different spheres of influence because they understand that institutions reflect, organize, and reproduce uneven power relations. This conflict often leads to changes in the institution or policy under consideration (Andreou, 2018).

Promoting Sustainable and Climate-Friendly Practices

A supportive environment is required for agricultural operations in order to use natural resources, generate agro-food, and maintain farmers' financial stability. Agriculture benefits the entire society, far beyond sustaining farming households and communities in rural areas (Birkhaeuser et al., 1991). However, has a twofold impact on the environment. Firstly, it directly impacts agricultural practices. Secondly, it contributes to climate change by releasing greenhouse gases into the atmosphere. To create a sustainable agricultural system throughout the EU, the CAP integrates social, economic, and environmental concerns. The EU has committed to additional international agreements to address climate change and sustainable development challenges (see also Vardopoulos & Karytsas, 2019), building upon a more innovative, more impactful, more comprehensive in its solutions to the challenges of climate change and sustainable development, and more ambitious and forward-thinking framework for environmentally friendly action. The original principles of the CAP did not prioritize environmental preservation and conservation, but this perception shifted as environmental issues became more politicized in the early 1970s. During the 1980s, several publications were released that highlighted the significance of safeguarding the environment. These include the "Green Paper" on the future of the CAP, a 1988 Communication on "Environment and Agriculture," and the guidebook "The Future of Rural Society." This literature emphasized the urgency of minimizing environmental degradation (Louloudis et al., 1999). The Green Paper report identified the importance of establishing institutionalized measures to mitigate and prevent environmental degradation caused by intensive farming practices (Vardopoulos et al., 2018). Techniques that prioritize reducing greenhouse gas emissions, storing carbon, and maintaining food production have the potential to alleviate the impacts of climate change (European Court of Auditors, 2021).

In the late 1980s and early 1990s, there was a surge in consumer and environmental activism that advocated for policy reform. This movement was prompted by several food-related scandals and the adverse ecological consequences of the agricultural practices supported by the CAP. Moreover, the EU intensified its international efforts to address environmental issues with global consequences (Carpenter, 2012), particularly after the United Nations Conference on the Environment and Development in Rio in 1992. These internal and external forces led to a major reform of the CAP in 1992, whereby environmental concerns became increasingly critical in the subsequent revisions of the policy. The 1992 agri-environmental measures were novel and reflected the first substantial effort for the support of agriculture as a providing pillar of goods and services that enhance the environment. The notion of the "second pillar" for rural development, which was initially introduced in Agenda 2000, continued to develop and expand this concept (Maravegias, 1983; Maravegias et al., 2023). The rural development pillar of the CAP was included in the Agenda 2000 reform package, which emphasized safe agri-food products and environmental outcomes. This reform package aimed to strike a balance between the need for environmental conservation and providing direct incentives to producers. Member States were required to take adequate environmental safeguards while being granted flexibility in how they supported farmers alongside environmental measures. When Member States failed to comply with regulations, their support funding was reduced or revoked, and the unpaid amounts were redirected to their respective rural development programs (Doukas, 2018; Louloudis & Maraveyas, 1997). Additionally, Member States were incentivized to spend a portion of these funds on developing more environmentally friendly production techniques in the dairy and cattle industries and training farmers in ecologically friendly practices to assist forests with high ecological value and underserved areas. The measures put in place aimed to enhance the efforts toward environmental preservation and climate action by requiring Member States to develop complete national or regional programs that included environmental conservation among other rural activities.

In the 2003 Mid-Term Review (MTR) of the CAP, cross-compliance became a mandatory requirement for all direct payments. The cross-compliance criteria are designed to ensure that farmers meet environmental and other standards before becoming eligible for subsidies. The regulations for statutory management under Union law and the requirements for maintaining excellent agricultural and environmental conditions were included in the cross-compliance norms (OECD, 2010). The MTR also introduced a "one farm payment" system, which is not based on output and is linked to compliance with environmental, food safety, and animal welfare criteria, as well as the obligation to maintain all farms in excellent agricultural and environmental condition. To promote the environment, quality, or animal welfare, direct payments were reduced for larger farms, freeing up more funds for programs that meet these goals (Cortignani et al., 2017; Maravegias & Martinos, 1997).

Over the last two decades and in order to maximize the use of natural resources, the CAP has urged farmers to adopt eco-friendly procedures for growing plants and raising animals as well as incorporating new technology into their production processes. Farmers that satisfy three environmental criteria are eligible for the green direct payment, which makes up to 30% of the direct payment program budget under the CAP system, diversifying their crops, preserving permanent grassland, maintaining biodiversity, and dedicating 5% of their arable land to ecologically beneficial areas (Ecological Focus Areas). In EU nations, the ratio of permanent grassland to agricultural land is decided by national or regional authorities, with a 5% leeway (Doukas, 2014).

Several studies have suggested that cross-compliance is an effective tool for promoting environmental and agricultural sustainability (Bennett et al., 2006; Juntti, 2012; Meyer et al., 2014; Ragazou et al., 2022). However, some scholars have argued that cross-compliance alone may not be sufficient to meet the environmental goals of the CAP and that it should be combined with other policy instruments, such as agri-environmental measures and payments for ecosystem services (Matthews, 2013; Meyer et al., 2014). Additionally, recent research has shown that the effectiveness of the CAP's green direct payment scheme may vary depending on the farm characteristics and the local context (Hristov et al., 2020), suggesting the need for more targeted and flexible policies (Doukas et al., 2023).

Therefore, policymakers and researchers need to continue examining the effectiveness of the CAP's environmental policies and identifying ways to improve them. The integration of agri-environmental measures, payments for ecosystem services, and the use of targeted policies based on farm and local context could be promising strategies to promote environmental sustainability while supporting agricultural production (Doukas & Petides, 2021).

In the EU, regulations are in place to protect designated sections of permanent grassland, which cannot be transformed or cultivated by farmers. However, certain farmers, such as those enrolled in the small farmer's program, are exempt from the greening regulations, and organic farmers are immediately rewarded for their environmentally friendly practices. Failure to comply with greening regulations can result in reduced direct payments. The green direct payment system is crucial in promoting good environmental practices and mitigating the impact of climate change on agriculture. The cross-compliance regime, although relatively lenient, provides a framework for regulation and control mechanisms. The agricultural industry worldwide faces increasing pressure from a growing population, urbanization, resource depletion, and climate change. In the EU, climate change effects, including extreme weather events, rising temperatures, and changing rainfall patterns, are adversely affecting agricultural production and environments, particularly in the southern and southeast regions. European agriculture is at risk of river flooding, droughts, coastal flooding, and other severe impacts due to climate change. While some regions may benefit from certain climatic changes, most will experience negative effects, exacerbating existing environmental issues.

Studies have shown that climate change is likely to increase the occurrence of extreme weather events, such as droughts and floods, with potentially serious effects on agricultural production (Doukas, 2019). The risks posed by climate change are particularly severe in the EU's southern and southeast regions, where agriculture is expected to be most negatively impacted (ibid). Furthermore, these effects on agriculture will likely exacerbate existing environmental issues, such as resource depletion, and lead to increased vulnerability in various areas (ibid). Therefore, implementing measures to mitigate the impact of climate change, such as the green direct payment system, is crucial in promoting sustainable agricultural practices and preserving the environment.

In addition, research has highlighted the importance of establishing precise and quantifiable standards for the cross-compliance regime, given its relatively lax framework and flaws in control mechanisms (Doukas, 2014). Moreover, farmers who disregard greening regulations are penalized through reduced direct payments, indicating the importance of complying with these regulations (ibid). Finally, the exemptions for certain farmers, such as those enrolled in the small farmer's program and organic farmers, should be carefully considered for administrative and proportionality reasons (Rydén, 2007).

Overall, protecting the environment and mitigating the impact of climate change are essential in maintaining sustainable agricultural production in the EU and globally. The implementation of regulations, such as the green direct payment system and the cross-compliance regime, is crucial in promoting good environmental practices and ensuring compliance with environmental standards.

To address the challenges facing agriculture from climate change and natural resource depletion, it is essential to implement sustainable production techniques and promote climate change and natural resource management. The CAP aims to achieve this goal since it acknowledges that farmers in Europe are the major environmental managers, spending money and producing in rural parts of Europe. The new CAP framework for 2021–2027 aims to promote a competitive and sustainable agricultural sector, supporting farmers' livelihoods and providing nutritious food to society while fostering vibrant rural communities (European Commission, n.d.). The European Green Deal, with a focus on agriculture and rural areas, is a crucial tool in achieving the Farm to Fork and biodiversity objectives (Fig. 2).

The new CAP requires measurable environmental and climatic requirements to be met to receive direct payments (Volkov & Melnikienė, 2017). This involves

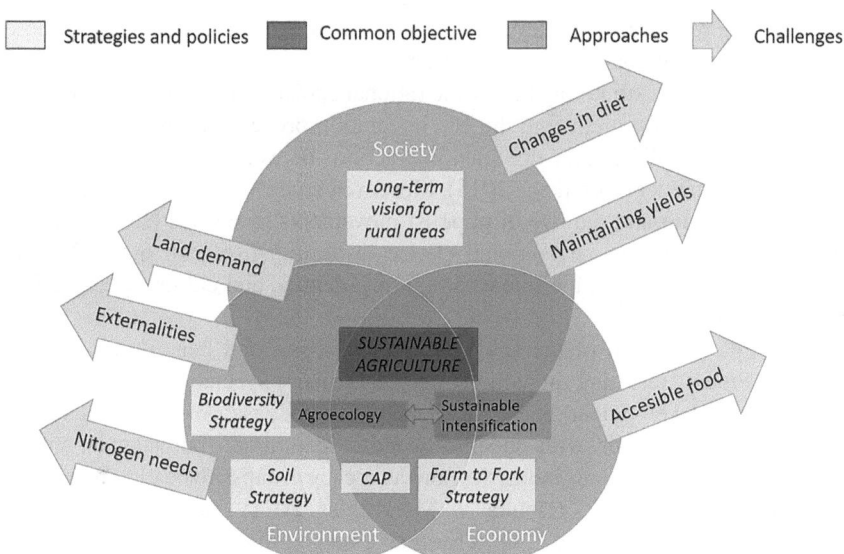

Fig. 2 Challenges and prospective approaches for achieving sustainable agriculture within the framework of the European Green Deal. (Source: Boix-Fayos & de Vente, 2023)

implementing agricultural practices such as the rotation of crops rather than diversification, safeguarding wetlands to maintain soils rich in carbon, and managing water resources in a sustainable manner. Each Member State must develop programs or incentives for farmers to encourage sound agricultural practices. The new CAP also includes an increased transfer of resources from pillar 1 to pillar 2 for environmental and climatic policies, reflecting a 15% financial upgradation of environmental issues (Bielik et al., 2022; Doukas et al., 2022).

Environmental Conditionality in the CAP: A Path toward Sustainable Agriculture

Historical institutionalism and rational choice institutionalism are two research methods used to explain the actions of Member State governments and their impact on the EU institutions. Leading scholars in the field of historical institutionalism, such as Pierson (1996) and Bulmer (2009), have emphasized the importance of examining political and policy-related activities in the context of multilevel government. Bulmer introduced the concept of a "governance regime" to analyze various subsystem policies within the EU.

On the other hand, rational choice institutionalism is based on the theory of rational behavior in economics and is primarily used to describe the goals of Member State governments in the EU integration process. As per this theory, governments willingly engage in and transfer power to the European Union because they believe it offers them several advantages. These benefits may include reduced transaction expenses, more effective policy formulation, improved efficiency, and increased adherence to regulation (Moravcsik, 1993).

However, it is worth noting that while rational choice institutionalism provides a useful framework for analyzing the EU integration process, it has also been criticized for oversimplifying the complex interactions between actors and institutions in the EU (Schneider & Ershova, 2018). Moreover, recent developments such as the Brexit referendum and the rise of populist movements in Europe have challenged the assumptions underlying rational choice institutionalism and highlighted the importance of examining the role of emotions, identity, and culture in EU politics (Manners, 2018).

The implementation of the CAP has varying costs and benefits for different Member States, resulting in competing goals and expectations among actors involved in its development. These entities may comprise various stakeholders, such as government officials with specific agendas, committee bureaucrats, interest groups representing various industries, and advocacy groups advocating for agricultural issues, among others (Doukas & Maravegias, 2021; Maravegias, 1991). By utilizing decisions regarding CAP reform as a bargaining tool, those in charge of agricultural policy have historically prioritized the concerns of farmers within their countries and those of their major trading partners. However, these decisions have often disregarded the opinions of consumer groups, especially in regard to environmental issues (Swinnen, 2001).

During the late 1990s, there was a notable transformation in the history of the CAP. This shift was triggered by several incidents. Examples of such incidents include the emergence of bovine spongiform encephalopathy, the detection of dioxin contamination in Belgian food products, and the spread of foot and mouth disease. These events brought about a heightened level of awareness among consumers, prompting them to demand greater attention to food safety and quality issues. This change in consumer demand also brought to light apprehensions regarding the influence of the CAP on animal welfare and the environment, leading to a push for the adoption of policies that address these issues. As a result, the Green Party leaders attempted to attend meetings of the appropriate ministers in countries such as Germany and Italy, which aided in the review and redefinition of the CAP (Swinnen, 2001).

The priority placed on food safety and quality has significantly risen among EU residents, largely due to the demand from consumer groups. Consequently, these issues have become a central focus of the CAP's agenda. These concerns have been on the rise, especially after the aforementioned situations, and continue to be a significant focus of the CAP. While the CAP's position on these issues may vary, its political response has been to address them in policy reform efforts (Swinnen, 2001).

Recent studies have emphasized the importance of aligning CAP policies with the EU's climate and biodiversity targets (Dupraz & Guyomard, 2019; Pe'er et al., 2020) and have highlighted the need to integrate environmental and climate objectives into the CAP to ensure its contribution to the EU's Green Deal objectives (Cuadros-Casanova et al., 2023; Pe'er et al., 2019). Scholarly research has emphasized the necessity of aligning the CAP with the Sustainable Development Goals established by the United Nations. Specifically, Goal 2, which aims to terminate hunger, attain food security, enhance nutrition, and foster sustainable agriculture, is of critical significance (Matthews, 2020; Swaminathan & Kesavan, 2018).

The CAP, established in 1962 and lasting until 1992, saw little opposition from pressure groups representing farmers' professional associations. However, in the following years, the European agricultural model's negative environmental impacts strengthened consumer and environmental movements. These movements subsequently gained power within EU institutions, due to food scandals and escalating environmental effects (Doukas, 2018).

Although politically justified, the economic philosophy behind cross-compliance, which involves adding new conditions to existing income support, remains unclear. The question arises whether this is a valuable policy tool. Several studies have indicated that direct payments were assigned based on agricultural policy objectives, rather than environmental objectives. As a result, farmers who heavily rely on direct payments may not necessarily align with those who cause significant harm to the environment (Doukas, 2018).

It is crucial to consider the decision-making process regarding the distribution of direct payment savings, particularly from farmers' noncompliance (Sadłowski et al., 2022). Member States could only withhold 25% of funds obtained through the implementation of cross-compliance (see Article 100 of the Regulation (EU) No 1306/2013), and therefore there was little incentive for states to establish a reliable control system. This lack of follow-through by the Member States indicates a discrepancy between political rhetoric and reality. Consequently, new "players" such as consumer and environmental movements gained importance in the reform process, as well as the gradual decline of producer pressure organizations resulting from the decline in the rural population, both in terms of absolute numbers and as a share of the overall workforce in the EU (Brady et al., 2017; Maravegias & Doukas, 2011).

It is worth noting that, historically, CAP's system perpetuates socio-economic disparities, particularly in rural areas, by favoring large farms and consolidating land ownership (Milczarek-Andrzejewska et al., 2018). This concentration of land ownership contributes to the displacement of small farmers and the exclusion of marginalized groups, perpetuating social injustice. The counterargument to this criticism is that direct payments are as concentrated as agricultural land, as 20% of the largest farms in the EU hold 82% of agricultural land and production (Matthews, 2020). The level of support is also determined by the specific products cultivated in specific areas. The products of the northern regions of the EU (e.g., grains, dairy) traditionally receive a higher support than those of the southern regions (e.g., wine, fruit, and vegetables), explaining the significant variation in the amount of financial aid received by each region of the EU (Shucksmith et al., 2005). On the other hand, the southern Member States with a large agricultural sector are primarily "poorer," and their farmers are small and less efficient; even if they produce products with low protection, the funding of the CAP is an essential proportion of their total income (Doukas & Maravegias, 2021). However, this fact does not compensate for an injustice that contradicts some of the policy's primary goals. Moreover, studies show that CAP has failed to deliver on its environmental objectives, despite a budget allocation for environmental purposes (Dupraz & Guyomard, 2019), indicating a need for reform to ensure sustainable agricultural practices.

The Green Payment Scheme represents a critical juncture in the evolution of the CAP in terms of both its design and implementation. With the introduction of this scheme, the compliance of producers with environmental regulations became quantifiable, and clear criteria were established for the allocation of direct payments. Furthermore, the most recent version of the CAP, covering the period from 2021 to 2027, places a greater emphasis on the targeted allocation of resources in order to align with the European Union's climate goals. This change is prompted by the recognition of the severe and widespread impacts of climate change on agriculture, which affects the entire food production chain. The new framework of the CAP aims to address these issues by implementing measures that specifically target climate action and sustainability in agriculture.

As demonstrated by the above examples, the focus on developing green architecture within the CAP is clear, aiming to create a robust yet adaptable framework for "greening" the policy. The CAP has been under scrutiny for a considerable period due to the detrimental effect it has on the environment. The increase in agricultural production intensity and the depletion of natural resources are among the factors causing this concern. Under the new CAP, rational decision-making is paramount within a structured framework of shared commitments and goals. The policy's diminished budget places considerable emphasis on conditionality, underscoring its significance. At the same time, the policy aims to raise awareness of and address environmental and climate change-related issues through specific funding mechanisms. Achieving these goals requires the integration of both CAP pillars, as well as cooperation with other related policies while providing flexibility based on national priorities (Doukas & Maravegias, 2021).

The Green Payment Scheme represents a new system of incentives aimed at promoting environmentally friendly agricultural practices, as well as rewarding farmers who have been meeting certain environmental standards. This is in line with the "public money for public goods" principle, which suggests that public funding for agriculture should be given in exchange for the provision of public goods, such as the protection of the environment and the preservation of biodiversity. The scheme is also designed to encourage farmers to take advantage of new technologies and management practices that reduce the environmental impact of agricultural production.

The new CAP framework for 2021–2027 builds on the previous policy framework, incorporating changes that reflect the EU's increasing emphasis on environmental sustainability and climate action (Matthews, 2018). In order to achieve this, the CAP encourages a competitive and sustainable agricultural sector supporting farmers' livelihoods while supplying society with nutritious, sustainable food and vibrant rural communities (Recanati et al., 2019). Sustainable farming practices include precision agriculture, organic farming, agroecology, agroforestry, and more stringent animal welfare standards (Doukas et al., 2022). By shifting the emphasis from compliance to performance, eco-schemes, for instance, should reward farmers for improved environmental and climatic performance, such as managing and storing carbon in the soil and improving fertilizer management to improve water quality and reduce emissions (Arata & Sckokai, 2016; Zafeiriou et al., 2023). The European Commission has recommended that climate action get at least 40% of the total funding for the CAP from 2021 to 2027 (Pe'er et al., 2020). Therefore, the Farm to Fork Strategy (European Commission, 2020) will support European farmers' efforts to combat climate change, protect the environment, and preserve biodiversity.

The European Green Deal is centered on agriculture and rural areas, and the new CAP aims to be a crucial tool in achieving the Farm to Fork and biodiversity targets. The policy's implementation relies on the combination of mandatory and voluntary measures, which vary according to the country's environmental and agricultural needs. The flexibility of the policy enables Member States to tailor their implementation strategies, depending on their individual circumstances, while the conditionality of direct payments serves as an incentive for farmers to meet environmental and climate-related objectives.

The adoption of the new CAP framework and the introduction of the Green Payment Scheme mark a noteworthy transition toward a more sustainable and environmentally aware approach to agriculture in the European Union. By incentivizing environmentally friendly practices and incorporating flexibility and conditionality, the policy establishes a foundation for integrating environmental considerations into the wider agricultural sector. Nonetheless, the effective implementation of this framework and the degree of acceptance and adaptation by Member States and farmers will ultimately determine its success in aligning with evolving policy priorities.

Concluding Notes

Over the last two decades, there has been a greater emphasis on policy change and adaptation to address the impact of economic activity and agricultural output on the environment and climate change. This change has been driven by growing awareness of the issue among consumers and the public, as well as recognition of agriculture's vulnerability to climate change due to the direct impact of weather on farming activities and its contribution to greenhouse gas emissions.

The new CAP for the programming period 2021–2027 represents a significant shift toward a more sustainable and environmentally friendly approach. The CAP's new green architecture integrates environmental and climate criteria that are quantifiable. The inclusion of wetland preservation for carbon-rich soils, responsible management of water resources, and crop rotation in lieu of diversification are some of the measures taken to ensure sustainability. In addition, the CAP allocates at least 30% of funding for environmental and climate change activities under the second pillar, amounting to roughly 23 billion euros.

The new CAP reflects a shared commitment to addressing climate change and sustainable development challenges and is an important step toward achieving these goals. It encourages collaboration with other similar policies, incorporates both CAP pillars, and increases flexibility based on national priorities. The new architecture represents a solid yet adaptable framework intended to "green" the CAP and to address the environmental and climate change concerns that have been raised for many years.

Moreover, alongside the green transition, the restoration of chronic injustices in the distribution of funding from the CAP should also be served. That could be achieved if the financial support stops following the unequal distribution of the agricultural area and volume of production between the agricultural holdings in the EU but favors more equitable funding, considering the specific income and development criteria in each region.

In conclusion, the new CAP framework and the Green Payment Scheme represent a significant shift toward a more sustainable and environmentally conscious approach to agriculture in the EU. The policy's focus on incentivizing environmentally friendly practices, combined with flexibility and conditionality, provides a comprehensive framework for the integration of environmental concerns and considerations into the broader agricultural sector. However, the success of the new framework relies on its effective implementation and the extent to which Member States and farmers embrace the new system of incentives and adapt to the policy's evolving priorities.

References

Alaimo, L. S., Ciommi, M., Vardopoulos, I., Nosova, B., & Salvati, L. (2022). The medium-term impact of the COVID-19 pandemic on population dynamics: The case of Italy. *Sustainability, 14*(21), 13995. https://doi.org/10.3390/su142113995

Andreou, G. (2018). *Η νεοθεσμική προσέγγιση της πολιτικής*. [The neo-institutional approach to politics]. Kritiki.

Arata, L., & Sckokai, P. (2016). The impact of Agri-environmental schemes on farm performance in five E.U. Member states: A DID-matching approach. *Land Economics, 92*(1), 167–186. https://doi.org/10.3368/le.92.1.167

Barasa, E., Mbau, R., & Gilson, L. (2018). What is resilience and how can it be nurtured? A systematic review of empirical literature on organizational resilience. *International Journal of Health Policy and Management, 7*(6), 491–503. https://doi.org/10.15171/ijhpm.2018.06

Beckert, J. (1999). Agency, entrepreneurs, and institutional change. The role of strategic choice and institutionalized practices in organizations. *Organization Studies, 20*(5), 777–799. https://doi.org/10.1177/0170840699205004

Bennett, C. J., & Howlett, M. (1992). The lessons of learning: Reconciling theories of policy learning and policy change. *Policy Sciences, 25*(3), 275–294. http://www.jstor.org/stable/4532260

Bennett, H., Osterburg, B., Nitsch, H., Kristensen, L., Primdahl, J., & Verschuur, G. (2006). Strengths and weaknesses of crosscompliance in the CAP. *EuroChoices, 5*(2), 50–57. https://doi.org/10.1111/j.1746-692X.2006.00034.x

Bielik, P., Turčeková, N., Adamičková, I., Belinská, S., & Bajusová, Z. (2022). Will changes in the common agricultural policy bring a respectful approach to environment in EU countries? *Visegrad Journal on Bioeconomy and Sustainable Development, 11*(1), 21–25. https://doi.org/10.2478/vjbsd-2022-0004

Birkhaeuser, D., Evenson, R. E., & Feder, G. (1991). The economic impact of agricultural extension: A review. *Economic Development and Cultural Change, 39*(3), 607–650. https://doi.org/10.1086/451893

Boix-Fayos, C., & de Vente, J. (2023). Challenges and potential pathways towards sustainable agriculture within the European green Deal. *Agricultural Systems, 207*, 103634. https://doi.org/10.1016/j.agsy.2023.103634

Bol, D., Giani, M., Blais, A., & Loewen, P. J. (2021). The effect of COVID-19 lockdowns on political support: Some good news for democracy? *European Journal of Political Research, 60*(2), 497–505. https://doi.org/10.1111/1475-6765.12401

Brady, M., Hristov, J., Höjgård, S., Jansson, T., Johansson, H., Larsson, C., et al. (2017). Impacts of direct payments – Lessons for CAP post-2020 from a quantitative analysis. .

Bulmer, S. (2009). Politics in time meets the politics of time: Historical institutionalism and the EU timescape. *Journal of European Public Policy, 16*(2), 307–324. https://doi.org/10.1080/13501760802589347

Capoccia, G. (2015). Critical junctures and institutional change. In *Advances in comparative-historical analysis* (pp. 147–179). Cambridge University Press. https://doi.org/10.1017/CBO9781316273104.007

Carmichael, D. G. (2015). Incorporating resilience through adaptability and flexibility. *Civil Engineering and Environmental Systems, 32*(1–2), 31–43. https://doi.org/10.1080/10286608.2015.1016921

Carpenter, A. (2012). Introduction: The role of the European Union as a global player in environmental governance. *Journal of Contemporary European Research, 8*(2). https://doi.org/10.30950/jcer.v8i2.533

Chantzaras, A., & Yfantopoulos, J. (2022). The impact of COVID-19 pandemic and its associations with government responses in Europe. *Region & Periphery, 13*(13), 23–40. https://doi.org/10.12681/rp.30758

Christiansen, T., & Verdun, A. (2020). Historical institutionalism in the study of european integration. In *Oxford research encyclopedia of politics*. Oxford University Press. https://doi.org/10.1093/acrefore/9780190228637.013.178

Clemens, E. S., & Cook, J. M. (1999). Politics and institutionalism: Explaining durability and change. *Annual Review of Sociology, 25*(1), 441–466. https://doi.org/10.1146/annurev. soc.25.1.441

Collier, R., & Collier, D. (1991). Critical junctures and historical legacies. In *Shaping the political arena: Critical junctures, the labor movement and regime dynamics in Latin America* (pp. 27–39). Princeton University Press.

Cortignani, R., Severini, S., & Dono, G. (2017). Complying with greening practices in the new CAP direct payments: An application on Italian specialized arable farms. *Land Use Policy, 61*, 265–275. https://doi.org/10.1016/j.landusepol.2016.11.026

Cuadros-Casanova, I., Cristiano, A., Biancolini, D., Cimatti, M., Sessa, A. A., Mendez Angarita, V. Y., et al. (2023). Opportunities and challenges for common agricultural policy reform to support the European green Deal. *Conservation Biology.* https://doi.org/10.1111/cobi.14052

Di Bartolomeo, G., Saltari, E., & Semmler, W. (2021). The effects of political short-termism on transitions induced by pollution regulations. In H. Dawid & J. Arifovic (Eds.), *Dynamic analysis in complex economic environments. Dynamic modeling and econometrics in economics and finance* (Vol. 26, pp. 109–122). Springer. https://doi.org/10.1007/978-3-030-52970-3_6

Doukas, Y. E. (2014). Ο αγροτικός τομέας την περίοδο 2014-2020. [The agricultural sector in the period 2014–2020]. In A. Mitsos (Ed.), *Οι πολιτικές που χρηματοδοτούνται από τον κοινοτικό προυπολογισμό και η ελληνική οικονομία.* [The Greek economy and the polices financed by the European Community]. Hellenic Foundation for European and Foreign Policy.

Doukas, Y. E. (2018). Η κοινή αγροτική πολιτική και ο ελληνικός αγροτικός τομέας. [The common agricultural policy and the Greek agricultural sector]. In N. Maravegias & T. Sakellaropoulos (Eds.), *Ελλάδα και Ευρωπαϊκή Ενοποίηση: Η Ιστορία μιας πολυκύμαντης σχέσης.* [Greece and European integration: The history of a multi-various relationship 1962–2018] (pp. 1962–2018). Dionicos.

Doukas, Y. E. (2019). The common agricultural policy under the pressure of the new financial framework (2021–2027): Nationalization and adaptation. *Region & Periphery, 8*, 133–142. https://doi.org/10.12681/rp.21159

Doukas, Y. E., & Maravegias, N. (2021). *Ευρωπαϊκή αγροτική οικονομία και πολιτική: Μετασχηματισμοί και προκλήσεις προσαρμογής.* [European agricultural economy and policy: Transformations and adaptation challenges]. Kritiki.

Doukas, Y. E., & Petides, P. (2021). The common agricultural policy's green architecture and the united Nation's development goal for climate action: Policy change and adaptation. *Region & Periphery, 11*, 107–128. https://doi.org/10.12681/rp.27247

Doukas, Y. E., Maravegias, N., & Chrysomallidis, C. (2022). Digitalization in the EU agricultural sector: Seeking a european policy response. In K. Mattas, G. Baourakis, C. Zopounidis, & C. Staboulis (Eds.), *Food policy modelling responses to current issues* (pp. 83–98). Springer. https://doi.org/10.1007/978-3-031-08317-4_6

Doukas, Y. E., Salvati, L., & Vardopoulos, I. (2023). Unraveling the European agricultural policy sustainable development trajectory. *Land, 12*(9), 1749. https://doi.org/10.3390/land12091749

Dupraz, P., & Guyomard, H. (2019). Environment and climate in the common agricultural policy. *EuroChoices, 18*(1), 18–25. https://doi.org/10.1111/1746-692X.12219

European Commission. (2020). Communication from the commission to the European Parliament, the council, the European economic and social committee and the Committee of the Regions. A farm to fork strategy for a fair, healthy and environmentally-friendly food system. COM(2020)381 Fina. European Commission. https://eur-lex.europa.eu/legal-content/EN/TXT/?uri=CELEX:52020DC0381

European Commission. (2021). CAP post-2020: Environmental benefits and simplification. https://agriculture.ec.europa.eu/system/files/2021-01/cap-post-2020-environ-benefits-simplification_en_0.pdf

European Commission. (n.d.). Key policy objectives of the CAP. https://agriculture.ec.europa.eu/common-agricultural-policy/cap-overview/cap-2023-27/key-policy-objectives-cap-2023-27_en#nineobjectives

European Court of Auditors. (2021). The common agricultural policy and climate change: Ambitious or inadequate? Special report 16/2021. https://www.eca.europa.eu/Lists/ ECADocuments/SR21_16/SR_CAP-and-Climate_EN.pdf

Greener, I. (2002). Understanding NHS reform: The policy-transfer, social learning, and path-dependency perspectives. *Governance, 15*(2), 161–183. https://doi.org/10.1111/1468-0491.00184

Hall, P. A. (1993). Policy paradigms, social learning, and the state: The case of economic policy-making in Britain. *Comparative Politics, 25*(3), 275. https://doi.org/10.2307/422246

Hall, P. A., & Taylor, R. C. R. (1996). Political science and the three new institutionalisms. *Political Studies, 44*(5), 936–957. https://doi.org/10.1111/j.1467-9248.1996.tb00343.x

Holley, K. A. (2009). Interdisciplinary strategies as transformative change in higher education. *Innovative Higher Education, 34*(5), 331–344. https://doi.org/10.1007/s10755-009-9121-4

Hristov, J., Clough, Y., Sahlin, U., Smith, H. G., Stjernman, M., Olsson, O., et al. (2020). Impacts of the EU's common agricultural policy "greening" reform on agricultural development, biodiversity, and ecosystem services. *Applied Economic Perspectives and Policy, 42*(4), 716–738. https://doi.org/10.1002/aepp.13037

Hu, X., & Yang, C. (2019). Institutional change and divergent economic resilience: Path development of two resource-depleted cities in China. *Urban Studies, 56*(16), 3466–3485. https://doi.org/10.1177/0042098018817223

Juntti, M. (2012). Implementing cross compliance for agriculture in the EU: Relational agency, power and action in different socio-material contexts. *Sociologia Ruralis, 52*(3), 294–310. https://doi.org/10.1111/j.1467-9523.2012.00564.x

Katsikas, D. (2022). Health and economic responses to the COVID-19 crisis in the EU. *Region & Periphery, 13*(13), 9–21. https://doi.org/10.12681/rp.30759

Kraatz, M. S., & Zajac, E. J. (1996). Exploring the limits of the new institutionalism: The causes and consequences of illegitimate organizational change. *American Sociological Review, 61*(5), 812. https://doi.org/10.2307/2096455

Lengnick-Hall, C. A., Beck, T. E., & Lengnick-Hall, M. L. (2011). Developing a capacity for organizational resilience through strategic human resource management. *Human Resource Management Review, 21*(3), 243–255. https://doi.org/10.1016/j.hrmr.2010.07.001

Levi, M. (1997). A model, a method, and a map: Rational choice in comparative and historical analysis. In M. Lichbach & A. Zuckerman (Eds.), *Comparative politics: Rationality, culture and structure* (pp. 19–41). Cambridge University Press.

Louloudis, L., & Maraveyas, N. (1997). Farmers and agricultural policy in Greece since the accession to the European Union. *Sociologia Ruralis, 37*(2), 270–286. https://doi.org/10.1111/j.1467-9523.1997.tb00050.x

Louloudis, L., Beopoulos, N., & Vlachos, G. (1999). Policy for the protection of the rural environment in Greece with a horizon in 2010. In N. Maravegias (Ed.), *Η ελληνική γεωργία προς το 2010.* [Greek agriculture towards 2010]. Papazisis.

Manners, I. (2018). Political psychology of european integration: The (re)production of identity and difference in the Brexit debate. *Political Psychology, 39*(6), 1213–1232. https://doi.org/10.1111/pops.12545

Maravegias, N. (1983). *Croissance economique et espace rural. Agriculture et agro-industrie dans le development regional de la Grece.* [Economic growth and rural space. Agriculture and agro-industry in the regional development of Greece]. Universite des Sciences Sociales de Grenoble, Faculte des Sciences Economiques.

Maravegias, N. (1991). Agriculture méditerranéenne et politique agricole commune: l'expérience de la Grèce. [Mediterranean agriculture and common agricultural policy: the experience of Greece]. In S. Bedrani & P. Campagne (Eds.), *Choix technologiques, risques et sécurité dans les agricultures méditerranéennes.* Montpellier. [Mediterranean agriculture and common agricultural policy: the experience of Greece] (pp. 159–165). CIHEAM.

Maravegias, N., & Doukas, Y. E. (2011). Traceability and the new CAP. In G. Baourakis, K. Mattas, C. Zopounidis, & G. van Dijk (Eds.), *A resilient european food industry in a challenging world. (European political, economic, and security issues; agriculture issues and policies)* (pp. xx–xx). Nova Science Publishers.

Maravegias, N., & Martinos, N. (1997). L'agriculture grecque face à l'union économique et moné-taire de la Communauté Européenne. [Greek agriculture facing the economic and monetary union of the European Community]. In A. Abaab, P. Campagne, M. Elloumi, A. Fragata, & L. Zagdouni (Eds.), *Agricultures familiales et politiques agricoles en Méditerranée : enjeux et perspectives*. [Family farming and agricultural policies in the Mediterranean: challenges and perspectives] (pp. 277–284). CIHEAM.

Maravegias, N., Doukas, Y. E., & Petides, P. (2023). The political economy of the common agricul-tural Policy's green architecture. *Sustainable Development, Culture, Traditions Journal, 1*(B), 73–84. https://doi.org/10.26341/issn.2241-4010-2023-1b-6

March, J. G., & Olsen, J. P. (1983). The new institutionalism: Organizational factors in political life. *American Political Science Review, 78*(3), 734–749. https://doi.org/10.2307/1961840

Matthews, A. (2013). Greening agricultural payments in the EU's common agricultural policy. *Bio-Based and Applied Economics, 2*(1), 1–27. https://doi.org/10.13128/BAE-12179

Matthews, A. (2018). The CAP in the 2021–2027 MFF negotiations. *Intereconomics, 53*(6), 306–311. https://doi.org/10.1007/s10272-018-0773-0

Matthews, A. (2020). The new CAP must be linked more closely to the UN sustainable development goals. *Agricultural and Food Economics, 8*(1), 19. https://doi.org/10.1186/s40100-020-00163-3

Meyer, J. W. (2017). Reflections on institutional theories of organizations. In *The SAGE handbook of organizational institutionalism* (pp. 831–852). SAGE. https://doi.org/10.4135/9781446280669.n33

Meyer, C., Matzdorf, B., Müller, K., & Schleyer, C. (2014). Cross compliance as payment for public goods? Understanding EU and US agricultural policies. *Ecological Economics, 107*, 185–194. https://doi.org/10.1016/j.ecolecon.2014.08.010

Milczarek-Andrzejewska, D., Zawalińska, K., & Czarnecki, A. (2018). Land-use conflicts and the common agricultural policy: Evidence from Poland. *Land Use Policy, 73*, 423–433. https://doi.org/10.1016/j.landusepol.2018.02.016

Moravcsik, A. (1993). Preferences and power in the european ommunity: A liberal intergovern-mentalist approach. *JCMS. Journal of Common Market Studies, 31*(4), 473–524. https://doi.org/10.1111/j.1468-5965.1993.tb00477.x

OECD. (2010). Environmental cross-compliance in agriculture. .

Ostrom, E. (1998). A behavioral approach to the rational choice theory of collective action: Presidential address, american political science association, 1997. *American Political Science Review, 92*(1), 1–22. https://doi.org/10.2307/2585925

Pe'er, G., Zinngrebe, Y., Moreira, F., Sirami, C., Schindler, S., Müller, R., et al. (2019). A greener path for the EU common agricultural policy. *Science, 365*(6452), 449–451. https://doi.org/10.1126/science.aax3146

Pe'er, G., Bonn, A., Bruelheide, H., Dieker, P., Eisenhauer, N., Feindt, P. H., et al. (2020). Action needed for the EU common agricultural policy to address sustainability challenges. *People and Nature, 2*(2), 305–316. https://doi.org/10.1002/pan3.10080

Pierson, P. (1996). The path to European integration. *Comparative Political Studies, 29*(2), 123–163. https://doi.org/10.1177/0010414096029002001

Pierson, P. (2000). Increasing returns, path dependence, and the study of politics. *American Political Science Review, 94*(2), 251–267. https://doi.org/10.2307/2586011

Polishchuk, L. (2012). Misuse of institutions: Lessons from transition. In *Economies in transition* (pp. 172–193). Palgrave Macmillan UK. https://doi.org/10.1057/9780230361836_8

Ragazou, K., Garefalakis, A., Zafeiriou, E., & Passas, I. (2022). Agriculture 5.0: A new strategic management mode for a cut cost and an energy efficient agriculture sector. *Energies, 15*(9), 3113. https://doi.org/10.3390/en15093113

Recanati, F., Maughan, C., Pedrotti, M., Dembska, K., & Antonelli, M. (2019). Assessing the role of CAP for more sustainable and healthier food systems in Europe: A literature review. *Science of the Total Environment, 653*, 908–919. https://doi.org/10.1016/j.scitotenv.2018.10.377

Regulation (EU) No 1306/2013. (2013). Regulation (EU) No 1306/2013 of the European Parliament and of the Council on the financing, management and monitoring of the common agricultural policy and repealing Council Regulations (EEC) No 352/78, (EC) No 165/94,

(EC) No 2799/98, (EC) No 814/2000. *Official Journal of the European Union*. https://eur-lex.
europa.eu/LexUriServ/LexUriServ.do?uri=OJ:L:2013:347:0549:0607:EN:PDF

Romanelli, E., & Tushman, M. L. (1994). Organizational transformation as punctuated equilibrium: An empirical test. *Academy of Management Journal, 37*(5), 1141–1166. https://doi.
org/10.5465/256669

Rydén, R. (2007). Smallholders, organic farmers, and agricultural policy. *Scandinavian Journal of
History, 32*(1), 63–85. https://doi.org/10.1080/03468750601160004

Sadłowski, A., Beluhova-Uzunova, R., Popp, J., Atanasov, D., Ivanova, B., Shishkova, M., &
Hristov, K. (2022). Direct payments distribution between farmers in selected new EU member states. *Agris on-line Papers in Economics and Informatics, 14*(4), 97–107. https://doi.
org/10.7160/aol.2022.140408

Schneider, G., & Ershova, A. (2018). Rational choice institutionalism and european integration.
In *Oxford research encyclopedia of politics*. Oxford University Press. https://doi.org/10.1093/
acrefore/9780190228637.013.501

Scott, W. R. (2004). Reflections on a half-century of organizational sociology. *Annual Review of
Sociology, 30*(1), 1–21. https://doi.org/10.1146/annurev.soc.30.012703.110644

Shucksmith, M., Thomson, K., & Roberts, D. J. (2005). *The CAP and the regions: The territorial
impact of the common agricultural policy*. CABI.

Streeck, W., & Thelen, K. (2005). Introduction: Institutional change in advanced political economies. In W. Streeck & K. Thelen (Eds.), *Beyond continuity: Institutional change in advanced
political economies* (pp. 1–39). Oxford University Press.

Strom, K. (1990). A behavioral theory of competitive political parties. *American Journal of
Political Science, 34*(2), 565. https://doi.org/10.2307/2111461

Swaminathan, M. S., & Kesavan, P. C. (2018). Science for sustainable agriculture to achieve UN
SDG goal 2. *Current Science, 114*(08), 1585. https://doi.org/10.18520/cs/v114/i08/1585-1586

Swinnen, J. F. M. (2001). A Fischler reform of the common agricultural policy? Centre
for European Policy Studies, Working Do. https://www.ceps.eu/ceps-publications/
fischler-reform-common-agricultural-policy/

Thelen, K. (1999). Historical institutionalism in comparative politics. *Annual Review of Political
Science, 2*(1), 369–404. https://doi.org/10.1146/annurev.polisci.2.1.369

Thelen, K., & Steinmo, S. (1992). Historical institutionalism in contemporary politics. In
S. Steinmo, K. Thelen, & F. Longstreth (Eds.), *Structuring politics: Historical institutionalism
in comparative analysis* (pp. 1–32). Cambridge University Press.

Tina Dacin, M., Goodstein, J., & Richard Scott, W. (2002). Institutional theory and institutional
change: Introduction to the special research forum. *Academy of Management Journal, 45*(1),
45–56. https://doi.org/10.5465/amj.2002.6283388

Trebesch, C., Konradt, M., Ordoñez, G., & Herrera, H. (2020). The political consequences of the
Covid pandemic: Lessons from cross-country polling data. *VoxEU.org*. https://cepr.org/voxeu/
columns/political-consequences-covid-pandemic-lessons-cross-country-polling-data

Vardopoulos, I., & Karytsas, S. (2019). An exploratory path analysis of climate change effects
on tourism. *Sustainable Development, Culture, Traditions Journal*, 132–152. https://doi.
org/10.26341/issn.2241-4002-2019-sv-13

Vardopoulos, I., Falireas, S., Konstantopoulos, I., Kaliora, E., & Theodoropoulou, E. (2018).
Sustainability assessment of the Agri-environmental practices in Greece. Indicators' comparative study. *International Journal of Agricultural Resources, Governance and Ecology, 14*(4),
368. https://doi.org/10.1504/IJARGE.2018.10019355

Volkov, A., & Melnikienė, R. (2017). CAP direct payments system's linkage with environmental sustainability indicators. *Public Policy and Administration, 16*(2), 231–244. https://doi.
org/10.13165/VPA-17-16-2-05

Wamsler, C. (2017). Stakeholder involvement in strategic adaptation planning: Transdisciplinarity
and co-production at stake? *Environmental Science & Policy, 75*, 148–157. https://doi.
org/10.1016/j.envsci.2017.03.016

Zafeiriou, E., Azam, M., & Garefalakis, A. (2023). Exploring environmental – Economic performance linkages in EU agriculture: Evidence from a panel cointegration framework.
Management of Environmental Quality: An International Journal, 34(2), 469–491. https://doi.
org/10.1108/MEQ-06-2022-0174

Climate Change Concerns and the Role of Research and Innovation in the Agricultural Sector: The European Union Context

Napoleon Maravegias, Yannis E. Doukas, and Pavlos Petides

Introduction

Agriculture faces a challenging and distinct problem as a consequence of climate change. First, agriculture is especially vulnerable because it depends heavily on weather and climate. Higher temperatures, more unpredictable rainfall, invasive pests, and increased extreme weather events are already detrimental to the industry, and these effects will worsen as climate change progresses. In addition, agriculture itself is a significant contributor to global greenhouse gas emissions (GHG), both directly (through emissions from production on farms) and indirectly (through changes in land use brought on by agricultural expansion). *Agriculture, Forestry, and Other Land Use* (AFOLU) accounts for around one-fifth (22%) of all global anthropogenic greenhouse gas emissions. Farms' methane and nitrous oxide emissions account for 50% of this; the remaining 50% comes from *Land Use, Land-Use Change, and Forestry* (LULUCF)-related CO2 emissions (Verschuuren, 2022). Methane reduction is crucial for stabilizing climate change by the middle of the century because it has an exceptionally high short-term impact on temperatures. Agriculture will continue to produce more emissions if nothing is done, and as other industries decarbonize, the sector's percentage of overall emissions could rise.

N. Maravegias
Department of Political Science and Public Administration, University of Athens, Athens, Greece

Y. E. Doukas (✉)
Department of Agricultural Development, Agri-food and Natural Resources Management, University of Athens, Psachna, Greece
e-mail: jodoukas@pspa.uoa.gr

P. Petides
Research Center for Economic Policy, Governance and Development, University of Athens, Athens, Greece

© The Author(s) 2024
A. Ribeiro Hoffmann et al. (eds.), *Climate Change in Regional Perspective*,
United Nations University Series on Regionalism 27,
https://doi.org/10.1007/978-3-031-49329-4_9

However, agriculture has many chances to lower direct and indirect emissions. Furthermore, via carbon storage in biomass and soils, agriculture provides natural methods for removing CO2 from the environment. Additionally, productivity-boosting strategies can be used to accomplish this. According to OECD research, the industry could contribute to mitigation at a pace of 8 Gt CO2eq/year in 2050, equivalent to two-thirds of current AFOLU emissions, with a complete policy package combining global emissions taxes and carbon sequestration subsidies (Heyl et al., 2021). Deforestation and other emissions related to land-use change would account for 62% of this total, while soil carbon sequestration would account for 29% of it (Pe'er et al., 2020).

Even with this potential, agriculture needs to catch up to other industries regarding pledges and activities toward climate change. Only 16 OECD member nations and significant emerging market economies had established objectives for reducing emissions in the agricultural sector by the middle of 2022. Only a few nations utilize targeted subsidies to encourage mitigation, and agriculture is typically excluded from mitigation policies like carbon prices or similar restrictions. Even though agriculture receives much policy assistance, very little fosters innovation or aligns with climate change goals.

The shift to more resilient and sustainable agriculture is hampered in particular by the fall in the share of support for general services, which includes agricultural knowledge and innovation systems and infrastructure, over the past two decades, from 16% to 13% (Dupraz & Guyomard, 2019). Many nations' current subsidies for agricultural output can raise their GHG emissions significantly. Although OECD countries are paying increasing attention to adaptation, existing plans focus primarily on short- and medium-term solutions than on building the transformative ability required to adjust to significant and ongoing environmental changes.

Responding to the above conditions, the European Union (EU), in the context of the new Common Agricultural Policy (CAP), which is into effect in 2023, aims to foster an agricultural industry that is competitive, resilient, and able to support farmers' livelihoods while also supplying society with wholesome, resilient food and vibrant rural communities. The European Green Deal is centred on agriculture and rural areas, and the new CAP aims to be a crucial tool in achieving the Farm to Fork and biodiversity targets. According to the document "2030 Digital Compass: The European Way for the Digital Decade", digital technology has the potential to significantly contribute to the achievement of the European Green Deal's objectives because the adoption of digital technologies and data will support the transition to a climate-neutral, circular, and resilient economy (Doukas et al., 2022).

The EU has also committed to new international agreements, such as the UN-Paris Agreement, dealing with climate change and sustainable development concerns. The Paris Agreement builds on the UN Framework Convention on Climate Change (UNFCCC) by bringing all nations together in the fight to effectively reduce greenhouse gas emissions and strengthen national capacities to build resilience and respond to the effects of climate change, including by ensuring that developing countries receive adequate assistance (UN, 2021).

In this chapter, the issues related to the interconnection of the agricultural sector with climate change will be examined, and the joint efforts for the climate goals at the global level will be described. Lastly, the role of research and innovation (R&I) in the agricultural sector in achieving them will be highlighted, especially in the context of the EU and the new CAP.

As it can be observed in Fig. 1, the new CAP is organized around ten key objectives for the years 2023–2027. These objectives, emphasizing social, environmental, and economic concerns, served as the blueprint for how EU nations constructed their CAP Strategic Plans. It is worth mentioning that climate change action, environmental care, and the preservation of landscapes and biodiversity are included as top priorities for the next years, along with the fostering of knowledge and innovation.

Climate Change Effects on Agriculture

Due to its enormous scale and sensitivity to weather conditions, which have significant economic effects, agriculture is the most vulnerable to climate change. Variations in climatic events like temperature and rainfall substantially impact the production of crops. The impact of changing precipitation patterns, rising temperatures, and CO_2 fertilization differs depending on the crop, the area, and the degree of parameter change. It has been discovered that rising temperatures decrease yield; however, rising precipitation is likely to cancel out or lessen the effects of rising temperatures (Adams et al., 1998). As seen in Iran under climatic variables, crop productivity is influenced by crop type, climate scenario, and CO_2 fertilization effect (Karimi et al., 2018). In Cameroon, it has been discovered that a drop in

Fig. 1 Key policy objectives of the CAP. (Source: European Commission, n.d.)

precipitation or a temperature rise dramatically reduces farmers' net income. The low demand for Cameroon's agricultural exports due to this problem and bad policymaking has caused volatility in national income. The impact of climate change on agriculture output varies by region and irrigation method. Expansion of irrigated regions can boost crop production but can harm the ecosystem. By shortening their growing seasons, many crops will probably produce less. If both the temperate and tropical regions experience a rise of 2 °C, the total yield of wheat, rice, and maize is anticipated to decline (Challinor et al., 2014). Tropical regions are more affected by climate change overall because tropical crops are still closer to their high-temperature optimums and are, therefore, more susceptible to high-temperature stress during high temperatures.

Additionally, humid and warm environments are more likely to have insect pests and diseases. Other factors affecting agricultural yields include humidity, wind speed, temperature, and rainfall. Without considering these factors, it is possible to overestimate the cost of climate change. Since the turn of the century, extreme weather events have increased in frequency in the Netherlands, substantially impacting wheat yield. The severity of the yield drop in wheat was determined by the week an extreme weather event occurred (Powell & Reinhard, 2016). In most of the world's areas, it has been predicted that there will be more droughts shortly due to climate change and that by 2100, the area impacted by droughts will have increased from 15.4 to 44.0%. The region is considered to be the most vulnerable in Africa. Significant crop yields in drought-affected areas are predicted to decline by more than 50% by 2050 and nearly 90% by 2100 (Verschuuren, 2022). Crop yield declines can drive food costs and severely impact agriculture's well-being globally, with a 0.3% annual loss in potential global GDP by 2100 (Wreford et al., 2010). The agriculture industry in India may suffer due to the expected rise in temperature in the range of 2.33 °C to 4.78 °C, doubling of CO_2 concentration, and lengthening of heat waves (OECD, 2019). The average crop yield in sub-Saharan Africa is predicted to decline by 6–24% due to climate change. Additionally, it is anticipated that Solomon Islands' overall fish demand will outpace fish output by 2050, significantly impacting food security as per-capita consumption will decline (OECD, 2022).

Agriculture's Impact on Climate Change

After the energy sector, the AFOLU sector is the second-largest emitter of GHGs globally, and the AFOLU accounts for 18% of GHG emissions in 2019 (Aguirre-Villegas & Craig, 2022). This is different in the EU, partly because there is not any deforestation, which in other regions of the world is frequently linked to agriculture (Fig. 2).

A decline in cattle numbers brought on by changes in agricultural practices in Eastern Europe changes to the Common Agricultural Policy (CAP), and the effects of policies enacted following the EU Nitrates Directive were the leading causes of the initial 20% decrease in agricultural emissions, which were primarily methane

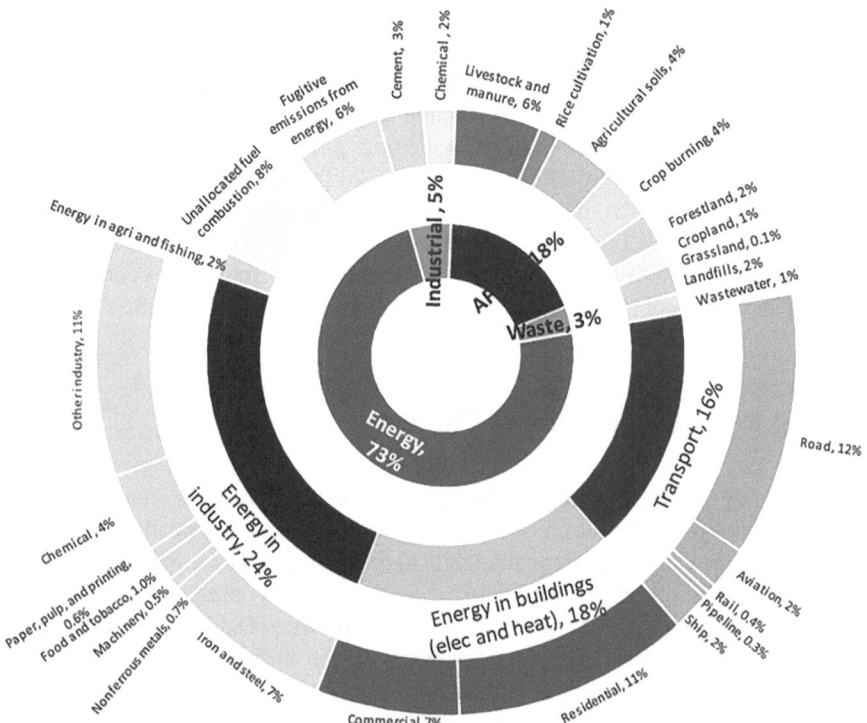

Fig. 2 Global GHG emissions by sector, where AFOLU is Agriculture, Forestry, and Other Land Use. (Source: Aguirre-Villegas & Craig, 2022)

and nitrous oxide emissions (Clark et al., 2020). However, emissions have significantly increased since 2014. Around 435 million metric tonnes of carbon dioxide (CO_2) equivalent was emitted in 2018, making up around 10% of all GHG emissions in the EU (Le Quéré et al., 2020). Approximately 70% of these come from cattle, 10 with the majority being enteric fermentation-related methane emissions, which account for 37% of all agricultural GHG emissions (Guerrero et al., 2022). Without any policy change, emissions are anticipated to stay in this range. To fulfill the EU's 2030 mitigation objective, it has been calculated that agricultural emissions must decrease by 25% by 2030 compared to 2015 (Huang, 2014). Reducing emissions from livestock, improving carbon storage (in agricultural soils and plants on fields), and restoring and managing peatlands are all ways to accomplish this.

It is worth mentioning that 9.3 billion tonnes of carbon dioxide equivalent of emissions connected to agriculture and related land use were produced globally in 2018. More than half of this total (5.3 Gt CO_2 eq) was produced by crop and animal activities within the farm gate, with land use and land-use change activities accounting for roughly 4 Gt CO_2 eq (Jenkins et al., 2018). In 2000, these parts were 4.6 and 5.0 Gt CO_2 eq, respectively. Emissions from the farm gate and land usage climbed during the 2000s, and trends in these two components started to diverge. Over the whole period of 2000–2018, emissions from agricultural and animal activities increased and were 14% higher in 2018 than they were in 2000 (Lynch et al., 2021). On the other hand, emissions from land use and land-use change declined across the study period, in line with trends in the amount of deforestation that were seen. As a result, the total agricultural emissions at the farm gate and on the field were around 4% lower in 2018 than in 2000 (EEA, 2022). In 2018, 17 percent of worldwide GHG emissions came from agriculture and related land use, down from 24 percent in the 2000s (EEA, 2022). This decline in emissions in 2018 resulted from the noted minor decrease in absolute emissions and the emissions from other economic sectors expanding at comparatively quicker rates between 2000 and 2018.

Global Response to Climate Change

Every part of the world is impacted by climate change. Sea levels are rising due to the melting of the polar ice caps. In certain areas, catastrophic weather events and flooding are growing more frequent, while high heat waves and droughts are getting more frequent in others. The official United Nations (UN) measures show that the average global temperature increased by 0.85 °C between 1880 and 2012. So, grain yields are decreased by about 5% for every degree the temperature rises. Corn, wheat, and other significant crops suffered significant output losses of up to 40 megatons per year between 1981 and 2002 due to a warmer environment. As seas grew owing to warming and glaciers melted, the average sea level rose globally between 1901 and 2010 by 19 cm. Global carbon dioxide (CO_2) emissions have increased by over 50% since 1990. Between 2000 and 2010, emissions rose more quickly than they had throughout the previous three decades. It will be more likely that global warming will not progress to this point if significant structural and technological reforms are made (UN, 2021).

The 2030 Agenda for Sustainable Development, which the EU28 and all other UN members endorsed in 2015, provides a shared framework for peace and prosperity for individuals and the global community. The 17 Sustainable Development Goals (SDGs) are at the core. Each goal typically has between 8 and 12 targets, and each target has 1–4 metrics that are used to monitor progress toward the goals. The objectives are either "outcome" targets (intended results) or "means of implementation" targets (UN, 2021). According to the Intergovernmental Panel on Climate Change (IPCC) in their 2018 Climate Report, limiting global warming to 1.5 degrees Celsius will need swift, extensive, and unmatched advances in all sectors of civilization. In Paragraph 14 of the Agenda, climate change is referred to as "one of the greatest challenges of our time", with fears that "its negative impacts endanger the willingness of all countries to advance sustainable development". The aims for Sustainable Development Goal 13 for Climate Change span various climate-related topics. There are a total of five targets. "Output targets" are the first three objectives including building knowledge and capacity to deal with climate change, increasing resilience and adaptive capacity to climate-related disasters, and incorporating climate change measures into policies and plans. Implementing the United Nations Framework Convention on Climate Change and promoting procedures to increase planning and management capacity are the final two objectives, which are "means of achieving" objectives. The leading worldwide intergovernmental platform for discussing the world's response to climate change is the UNFCCC. By addressing the dangers and possibilities climate change brings, Sustainable Development Goal 13 seeks to improve all nations' resilience and capacity for adaptation to climate-related hazards and natural disasters.

Global disruptions in human activity and development were brought about by the COVID-19 pandemic in 2020, with some of these changes having a favourable impact on GHG emissions. The use of coal-fired power plants was dramatically decreased, especially in China, due to a 5% decrease in domestic and international energy demand (UN, 2021). The EU27's proportion of global emissions declined from 16.8% in 1990 to 7.3% in 2021, a decrease of 27.3% from 1990 levels (Crippa et al., 2022). As a result, UNEP supports authorities and investors in funding fiscal stimulus plans and prioritizing green and decent employment. Emissions increased once the world economy recovered from the pandemic. CO_2 emissions increased by 5.3% in 2021 compared to 2020, totalling 37.9 Gt CO_2, just 0.36% less than in 2019. The top emitters of CO_2 worldwide were China, the United States, the EU27, India, Russia, and Japan. Together, they were responsible for 67.8% of the world's fossil CO_2 emissions, 66.4% of its fossil fuel consumption, and 49.2% of the world's population (UN, 2021). In 2021 compared to 2020, all six significant emitters increased their fossil CO_2 emissions, with India and Russia experiencing the most significant percentage increases (10.5% and 8.1%) (Crippa et al., 2022). As stated in the United Nations Framework Convention on Climate Change, nations should fulfill their commitments to fully put into practice the Green Climate Fund, tackle the needs of developing nations in the light of meaningful mitigation actions, and mobilize $100 billion annually from all sources by 2020. Focusing on women, youth, and local, underserved groups help countries promote methods for enhancing the ability of the least developed nations and small island developing states for effective climate change planning and management.

The Paris Agreement builds on the UNFCC 1, bringing all nations together in the fight to quickly reduce greenhouse gas emissions, strengthen national capacities, foster resilience, and respond to the effects of climate change, including by guaranteeing that developing nations receive adequate support. With the early entry into force of the Paris Agreement and the practical introduction of the Katowice Climate Package, the world has reached a new period in its collective efforts to combat climate change, concentrating on urgently growing commitment and implementation at all levels of government, industry, and civil society (UN, 2021).

Most of the worst effects of climate change are too severe and too quickly occurring for adaptation strategies to be effective, which has posed a new problem that has been a significant discussion point during the Paris negotiations. In particular, the Paris Agreement acknowledges the need to address losses and damages of this kind and aims to develop appropriate solutions. According to this statement, loss and destruction can occur in various ways, including rapid effects of severe weather and slow-rise effects, like a land loss at sea for lower islands with serious adverse effects on agricultural production (Climate Focus, 2016).

Although the NDCs of each Party may not be legally binding, the Parties are legally obligated to review their development toward the NDC and find ways to support their goals. In Article 13 of the Paris Accord, which establishes uniform criteria for monitoring, reporting, and verification (MRV), the phrase "enhanced accountability system for action and assistance" is used. As a result, both rich and developing countries must submit reports on their mitigation efforts every 2 years and be subject to technical and peer evaluation (Climate Focus, 2016). The Agreement recognizes the various circumstances of various countries and declares that competent expert evaluations appreciate the distinctive reporting capacities of each nation (Asselt, 2018). Accordingly, at the 2015 Paris Conference, when the Agreement was considered and determined to mobilize $100 billion in climate finance by 2025, the less developed countries reaffirmed their promises to mobilize $100 billion annually on climate financing by 2020. The money will be used to promote development mitigation and adaptation. The UNFCCC Green Climate Fund and several other public and private initiatives are funded using this money. The Paris Agreement requires a new lease of $100 trillion annually to be agreed upon through 2025 (Roberts et al., 2021). Parties at the UN Climate Change Conference (COP27), which concluded in Egypt on 2022, agreed that limiting global warming to 1.5 C required rapid, deep, and persistent reductions in global greenhouse gas emissions, with a 43 percent reduction by 2030 relative to the 2019 level. They underlined the need from the Glasgow Climate Pact for nationally determined contributions (NDCs) to be adjusted as necessary by the end of 2023 in order to align with the Paris Agreement temperature objective. Additionally, they reaffirmed that a new mitigation work programme will be guided by the Glasgow Climate Pact in order to urge Parties to align their goals and activities in the direction of net zero (Doukas & Petides, 2021). The Paris Agreement is a legally binding agreement that, as opposed to the Kyoto Protocol and the Copenhagen Accord, draws all countries collectively for the first time in the multilateral climate change process to carry out bold steps to tackle and accommodate climate change. The Paris Convention affirms that despite allowing initial volunteer contributions from other Parties, wealthy countries should take the lead in providing financial assistance to less compliant and needy nations. Because significant emissions reductions demand large-scale investments, mitigation necessitates climate financing. Since significant financial resources are needed to adapt to the harmful effects and lessen the effects of climate change, climate finance is also essential for adaptation. Nations fashioned a more open system with the Paris Agreement (ETF). In 2024, nations will provide transparent reports on their actions, their progress, and the support they have received or provided under the ETF. Additionally, it specifies global guidelines for examining articles that have been submitted (UN, 2021).

Despite the COVID-19 pandemic's marginally positive effects on pollution reduction, SDG 13 still has several obstacles to overcome. According to the 2022 report on CO2 emissions of all world countries, compiled by the JRC, the International Energy Agency (IEA), and the Netherlands Environmental Assessment

Agency, global fossil CO2 emissions rose by 5.3% in 2021 compared to 2020, approaching pre-pandemic 2019 levels (Crippa et al., 2022). Unless an emphasis is given on green agreements when transferring monetary money, financing economic policy would likely redirect emergency funds often devoted to environment funding, such as the Green Climate Fund and environmental policies. As government lockout measures are loosened, transportation emissions are expected to rise. Also, nations which experienced a decrease in their productivity levels tend to restrict compliance with environmental standards. Furthermore, the COP27 – held in Sharm El-Sheikh, Egypt, in November 2022 – failed to progress on commitments or prove that nations are willing to take significant action to reduce global emissions (World Economic Forum, 2022). Therefore, the world could not keep the global temperature increase below 1.5 degrees Celsius, a temperature goal set in the Paris Agreement.

SDGs (13) calls for "urgent action to tackle climate change and its impacts" because it is universally acknowledged as a threat that defines our time. 70% of studies on the effects of climate change predict declining crop yields by 2030, with half of those studies predicting decreases of between 10% and 50%. Climate change already impacts food systems, and agriculture is one of the most severely impacted industries. About 25% of global annual GHG emissions are caused by agriculture and related land-use shift. It would be necessary to significantly reduce emissions in the food systems if the global warming goal were not realized. As a result, many adaptation and mitigation measures in the food systems would be required to achieve SDG 13. The fact that food systems are linked to several SDGs and that food system behaviour may result in trade-offs across SDGs is a significant problem, with trade-offs being more challenging in developing nations where climate change vulnerability is highest (Doukas & Petides, 2021). The food system must change significantly to meet SDG 13 and UNFCCC commitments. However, this change must consider the possibility of trade-offs between other SDGs, such as adaptation and mitigation. The difficulties are so great that a complete revolution in food systems is required, with specific behaviour determined by context. There are various ways in which food systems are evolving. However, many academics contend that the shift needs to be considerably more significant for food security, climate change mitigation, and environmental sustainability in the coming years. The Food and Agricultural Organization (FAO) is additionally helping countries adapt to and mitigate the effects of climate change by creating national climate plans and putting into action research-based programmes and initiatives, focusing on smallholder agriculture and strengthening the livelihoods of rural communities (FAO, 2019).

Research and Innovation in the Agricultural Sector: Benefits and Risks

In times of successive economic, geopolitical, health, and climate crises, exploring the extent to which research and technology can be significant parameters in initiating a sustainable development process is essential. The role that agriculture, and more broadly the rural space, can play in this effort is worthy of investigation, given the vital need to produce sufficient, safe food to feed the world's population. The concept of sustainable agriculture has gained a central position in the public debate to mitigate climate change effects, among others, and new technologies are called upon to provide solutions that allow the achievement of purely economic goals (food sufficiency, productivity, and efficiency) on the one hand but also relating to the safeguarding of public health, environmental and climate protection, and social cohesion (Doukas & Maravegias, 2021; Labrianidis et al., 2005).

At the same time, it is worth noting that, throughout the agro-food chain, the human factor plays an equally important role, either from the producer's or the consumer's side. In particular, the farmer-producer tries to respond to the rapid changes taking place in the agricultural society but also in the markets of agricultural products and at the same time to fulfill his multiple roles as a producer of healthy and safe food products, a modern entrepreneur, and a central factor in the development of the rural space (Maravegias & Doukas, 2011). Therefore, the farmer should respond by increasing productivity, improving the quality of the produced product, and responding immediately to the demands of the market – by introducing new products into the production process, demonstrating adaptability but also the ability to immediately integrate new processes into the agricultural production (Siardos & Koutsouris, 2004).

Also, R&I in the agricultural sector is constantly intensifying and includes a wide range of applications in biotechnology, digital information technology, communications, production of new products, use of new inputs, and organic agriculture. These developments significantly affect agricultural production, as they play a decisive role in forming modern methods in agro-food sector and their subsequent effectiveness (Doukas & Maravegias, 2021).

By the nature of the agricultural process, technology applications are carried out locally, yet the production of these applications is highly internationalized and primarily concentrated in private companies (Schimmelpfennin & Thirtle, 1999). Therefore, the involvement of multinational corporations, organizations, and their subsidiaries is crucial in transferring and disseminating know-how. As mentioned, the local character (climate and physical factors) dominates agricultural production. So, their local subsidiaries produce research work to adapt these products to the needs of local production and the particular requirements of technology demand, as they are formed at a local level. As a result, every aspect of contemporary agriculture technology and information dissemination is currently being actively pursued by global corporations. It might cover everything from mechanical machinery breakthroughs to new propagation materials for producing plants and animals.

Utilizing contemporary agricultural technologies requires farmers to have high professional training (Apostolopoulos, 2004; Doukas & Maravegias, 2021).

Developing new organizational structures to combine the various factors in the production process through the integrated management of agricultural holdings is an essential outcome of agricultural R&I, based mainly on digital applications. Integrated management includes the best possible utilization of agronomic data in production, with weather sensors, drones, and GPS, to name a few. Also, economic and agricultural research shows that participatory production to achieve economies of scale by sharing high-cost fixed capital equipment contributes to better income outcomes. At the same time, environmental and climate goals are ensured, through the more efficient use of production factors and the subsequent reduction of greenhouse emissions.

On the other hand, modern trends in R&I and technology have brought to the fore issues that are the subject of intense reflection, such as the benefits and risks deriving from the use of genetically modified organisms, food safety issues, and environmental effects from the implementation of new production processes and methods, whose effects on the ecosystems cannot be immediately seen, nor can they be accurately predicted, significantly when the level of education of the farmers does not contribute to the evaluation, from their side, of the above critical issues (Doukas, 2018). Additionally, the modern organizational systems of the agro-food chain, which prioritize consumers' "needs and wants" while also intending to protect the environment, are characterized mainly by structures of monopolistic competition and market power concentration and require significant cutting-edge applications in the supply chain, which not all farmers can support based on their economic position. Thus, agriculture of different speeds is created worldwide and the economic inequalities between the participants in the global food systems widen.

Research and Innovation: The EU Context

Within the EU, the objectives of the new CAP for the period 2023–2027 include ensuring a fair income for farmers, action on climate change, encouraging generational renewal, increasing competitiveness, protecting the environment, developing dynamic rural areas, balancing power in the food chain, preservation of landscapes and biodiversity, protection of health, and food quality (EC, 2018a, b, 2020). Particular importance is given to "Green Architecture" or "Green Deal" for the agricultural sector's environmentally sustainable development. They also agreed that 30% of total EU Budget spending, including the COVID-19 Recovery Fund (Next Generation EU), should contribute to climate goals. In this direction, it was decided that 40% of the expenditure of the new CAP should be committed to achieving the above objectives (EC, 2020).

Based on the above, there are many challenges to which the new CAP must respond in the coming years. The most important ones include the economic strength and sustainability of the agricultural sector, ensuring the proper management of the natural environment, actions to tackle climate change, creating a solid and cohesive economic and social fabric in the EU's rural areas, and exploiting emerging opportunities for action in the fields of global trade, bioeconomy, renewable energy, and the digital economy (Maravegias et al., 2023). Incentives are also given for developing "smart" applications (precision agriculture, improvement of broadband connections) and developing a pan-European risk management platform. Finally, the strategic plan of each member-state should necessarily include actions for exchanging knowledge and innovation, a commitment that requires the modernization of the respective state services (Doukas, 2019).

On the world map of the production of know-how and innovations in the agricultural sector, the EU has been developing intense activity over the last three decades (e.g. the Horizon programme), while for the education and training of farmers, it has significant resources. Through various research programmes, particular emphasis is placed on food safety, animal health, and the environmental impact of agriculture, and at the same time, a strict policy of exclusion from the European internal market of genetically modified products is practised (Doukas & Maravegias, 2021). In addition, education, training, research, technology, and innovation actions with applications in the agricultural economy and rural development are financed mainly through the European Agricultural Fund for Rural Development (EAFRD) of the CAP. Also, research and technology issues are in an even more central position in the Commission's new proposals for the CAP concerning the period 2021–2027 (Doukas, 2019).

Also, the EU funds research initiatives that involve both public and private sector organizations and have a global scope. The relationship between research and professional education is also encouraged, and cooperation between productive and academic organizations is strengthened. The funding of the initiatives mentioned above is typically anticipated, with sums up to 10 billion euros within the framework of the Horizon programme (EC, 2018b). However, there is a delay in adopting new technology by farmer-producers in the agricultural sector since they need to become more familiar with innovations and new production systems due to their lack of agricultural education. Large groups of farmers experience erosion in their ability to compete and a squeeze in their income.

The established institutional framework and incentives offered for environmentally friendly agricultural practices and the adoption of animal welfare and euthanasia practices, both in the EU and internationally, are redefining the orientation of agricultural production and, to some extent, determining the direction of technological advancements in the agricultural sector. Therefore, with the primary goal of preserving natural resources and protecting the environment, the use of less-polluting procedures and methods, integrated systems for the utilization of agricultural waste, and recycling of valuable materials direct agricultural production toward products for niche markets, such as organic products and goods for energy purposes, and require, in several cases, new technological applications.

As mentioned above, climate change has brought significant disruption in primary crops; it is expected that technology applications and digitization of the agrofood sector will lead to the most efficient use of depleting natural resources with the lowest possible environmental footprint. So, to hasten the transition from primary production to consumption of sustainable, wholesome, and inclusive food systems, the Farm to Fork Strategy – the central pillar of the new CAP – acknowledges that R&I is essential driver. R&I may help with the creation and testing of solutions, the removal of barriers, and the identification of new market opportunities. The European Innovation Partnership's Agricultural Productivity and Sustainability (EIP-AGRI) will play a more significant part in the strategic plans of the member states to promote innovation and knowledge transfer. The Commission plans to work with the member states on this. The European Regional Development Fund will also contribute to the collaboration and innovation of the food value chain through smart specialization (Doukas et al., 2022).

They can also contribute to the sustainability of agricultural systems from an economic, social, and environmental point of view compatible with the EU's Green Deal and the Farm to Fork Strategy objectives to secure climate goals, the maintenance of biodiversity, and the European echo systems (EC, 2019). Such technologies can optimize all types of agriculture, facilitate better decision-making, and reshape the functioning of agricultural markets throughout the food chain, and are in line with the EU Commission's recognition that since the so-called twin transitions to a green and digital Europe continue to be the generation's defining issues, the CAP must be at the forefront of the shift to more sustainable and climate-neutral agriculture (Doukas et al., 2022).

Conclusions

In the threatening context of climate change, agriculture is in a complex and distinct circumstance as it is particularly vulnerable due to its reliance on weather and climate. As it was illustrated in this chapter, the effects of climate change on the industry are already adverse and will worsen as it progresses. These effects include rising temperatures, unpredictable rainfall, invasive pests, and increased extreme weather events. A considerable portion GHG emissions are also caused by agriculture itself, both directly (through emissions from farm production) and indirectly (through changes in land use brought on by agricultural expansion).

Therefore, investigating the extent to which R&I in the agricultural sector can address the above challenges is essential. Various applications in biotechnology, digital information technology, communications, new product development, new inputs, and organic agriculture are areas where agricultural R&I is continually intensifying. These changes substantially impact agricultural output since they are essential to developing contemporary agro-food sector methods and their success.

Nevertheless, there is a need for a digitally skilled workforce to support the modernization process of the agricultural sector. The potential for the spread of digital technology in European agriculture is much greater than in developing countries. However, the needs for digital technology are also more significant as the demands of the consumer and environmental movement in Europe are higher than in other parts of the world. Thus, advanced digital technologies such as artificial intelligence, robotics, and "5G" can improve the efficiency of farms and increase their productivity in the EU.

Due to the nature of the agricultural process, in order to transmit and spread knowledge, international firms, organizations, and their subsidiaries need to be involved. As a result, multinational firms are actively pursuing every facet of modern agriculture technology and information transmission. Due to their differing economic circumstances, not all farmers can sustain the numerous cutting-edge supply chain applications required by such monopolistic competition and market power concentration structures. Therefore, agriculture is developed at different speeds worldwide, and the economic disparities between the participants in the global food systems increase. This poses challenges to the EU and Latin America cooperation and should be addressed in the political dialogue between the EU, CELAC, and other Latin American regional organizations.

In the EU framework, "Green Architecture" is given a particular position for the agricultural sector's sustainable development in promoting the transition to a climate-neutral, circular, resilient rural economy. At the same time, the EU has also committed to new international agreements, such as the UN-Paris Agreement, dealing with climate change and sustainable development concerns. The Farm to Fork Strategy, which provides the core of the new CAP, recognizes that (R&I) constitutes an essential driver. Developing and testing solutions, removing obstacles, and discovering new market opportunities may all be aided through R&I. In order to effectively address the problems posed by climate change, the European Innovation Partnership's Agricultural Productivity and Sustainability (EIP-AGRI) is supposed to play a more significant part in the member states' strategic plans.

References

Adams, R. M., Hurd, B. H., Lenhart, S., & Leary, N. (1998). Effects of global climate change on agriculture: An interpretative review. *Climate Research, 11*(1), 19–30.

Aguirre-Villegas, H., & Craig, B. (2022). Expectations for coal demand in response to evolving carbon policy and climate change awareness. *Energies, 15*(10), 3739–3752. https://doi.org/10.3390/en15103739

Apostolopoulos, K. (2004). Το παρόν και το μέλλον της Ελληνικής Γεωργίας- Τεχνολογικές Πτυχές. In N. Maravegias (Ed.), *Στρατηγική για την Αγροτική Ανάπτυξη της Ελλάδας*. Papazisis. ISBN: 9600217777.

Asselt, H. (2018). Putting the 'enhanced transparency framework' into action. Stockholm Environment Institute, from https://www.sei.org/publications/enhanced-transparency-framework/

Challinor, A. J., Watson, J., Lobell, D. B., Howden, S. M., Smith, D. R., & Chhetri, N. A. (2014). Meta-analysis of crop yield under climate change and adaptation. *Nature Clim Change, 4*, 287–291. https://doi.org/10.1038/nclimate2153

Clark, M. A., Domingo, N. G. G., Colgan, K., Thakrar, S. K., Tilman, D., Lynch, J., et al. (2020). Global food system emissions could preclude achieving the 1.5° and 2°C climate change targets. *Science, 370*, 705–708. https://doi.org/10.1126/science.aba7357

Climate Focus. (2016). *COP21 Paris 2015 – climate focus overall summary and client briefs.* Climate Focus, from https://www.climatefocus.com/publica-tions/cop21-paris-2015-climate-focus-overall-summary-and-client-briefs

Crippa, M., et al. (2022). CO2 emissions of all world countries. In *Publications Office of the European Union.* https://doi.org/10.2760/730164

Doukas, Y. E. (2018). Η κοινή αγροτική πολιτική και ο ελληνικός αγροτικός τομέας. In N. Maravegias & T. Sakellaropoulos (Eds.), *Ελλάδα και ευρωπαϊκή ενοποίηση: Η ιστορία μιας πολυκύμαντης σχέσης 1962–2018.* Dionicos. ISBN: 9789606619823.

Doukas, Y. E. (2019). The common agricultural policy under the pressure of the new financial framework (2021–2027): Nationalisation and adaptation. *Region & Periphery, 8,* 133–142. https://doi.org/10.12681/rp.21159

Doukas, Y. E., & Maravegias, N. (2021). *Ευρωπαϊκή αγροτική οικονομία και πολιτική: Μετασχηματισμοί και προκλήσεις προσαρμογής.* Kritiki. ISBN: 9789605863753.

Doukas, Y. E., & Petides, P. (2021). The common agricultural Policy's green architecture and the united Nation's development goal for climate action: Policy change and adaptation. *Region & Periphery, 11,* 107–128. https://doi.org/10.12681/rp.27247

Doukas, Y. E., Maravegias, N., & Chrysomallidis, C. (2022). Digitalisation in the E.U. Agricultural sector: Seeking a european policy response. In K. Mattas, G. Baourakis, C. Zopounidis, & C. Staboulis (Eds.), *Food policy modelling responses to current issues* (pp. 83–98). Springer. https://doi.org/10.1007/978-3-031-08317-4_6

Dupraz, P., & Guyomard, H. (2019). Environment and climate in the common agricultural policy. *EuroChoices, 18,* 18–25. https://doi.org/10.1111/1746-692X.12219

European Commission. (2018a). *Proposal for regulation of the European Parliament and of the council,* COM (2018) 392 final, Brussels.

European Commission. (2018b). *A modern budget for a Union that protects, empowers and defends: The multiannual financial framework for 2021–2027,* COM (2018) 321 final, Brussels.

European Commission. (2019). The European green Deal. COM(2019) 640 final, Brussels.

European Commission. (2020). Statistical factsheet: European Union, Agriculture and Rural Development. https://ec.europa.eu/info/sites/info/files/food-farming-fisheries/farming/documents/agri-statistical-factsheet-eu_en.pdf

European Commission. (n.d.). Key policy objectives of the CAP. https://agriculture.ec.europa.eu/common-agricultural-policy/cap-overview/cap-2023-27/key-policy-objectives-cap-2023-27_en#nineobjectives

European Environment Agency. (2022). Greenhouse gas emissions from agriculture in Europe. EEA. https://www.eea.europa.eu/ims/greenhouse-gas-emissions-from-agriculture

FAO. (2019). FAO's work on climate change: United Nation's climate change conference. http://www.fao.org/3/ca7126en/ca7126en.pdf

Guerrero, S., et al. (2022). The impacts of agricultural trade and support policy reform on climate change adaptation and environmental performance: A model-based analysis. In *OECD food, agriculture and fisheries papers, no. 180.* OECD Publishing. https://doi.org/10.1787/520dd70d-en

Heyl, K., Döring, T., Garske, B., Stubenrauch, J., & Ekardt, F. (2021). The common agricultural policy beyond 2020: A critical review in light of global environmental goals. *RECIEL, 30,* 95–106. https://doi.org/10.1111/reel.12351

Huang, J. (2014). Climate change and agriculture: Impact and adaptation. *Journal of Integrative Agriculture, 13*(4), 657–659.

Jenkins, S., Millar, R. J., Leach, N., & Allen, M. R. (2018). Framing climate goals in terms of cumulative CO2-forcing-equivalent emissions. *Geophysical Research Letters, 45*(6), 2795–2804. https://doi.org/10.1002/2017GL076173

Karimi, V., Karimi, E., & Keshavarz, M. (2018). Climate change and agriculture: Impacts and adaptive responses in Iran. *Journal of Integrative Agriculture, 17*(1), 1–15. https://doi.org/10.1016/S2095-3119(17)61794-5

Labrianidis, L., Kalogeresis, T., & Kourtesis, A. (2005). "Νέες τεχνολογίες, καινοτομία και ανάπτυξη της υπαίθρου", στο. In L. Labrianidis (Ed.), *Η επιχειρηματικότητα στην ευρωπαϊκή ύπαιθρο: Η περίπτωση της Ελλάδας*. Patakis. ISBN: 139789601616896.

Le Quéré, C., Jackson, R. B., Jones, M. W., Smith, A. J. P., Abernethy, S., Andrew, R. M., et al. (2020). Temporary reduction in daily global CO2 emissions during the COVID-19 forced confinement. *Nature Climate Change, 10*, 647–653. https://doi.org/10.1038/s41558-020-0797-x

Lynch, J., Cain, M., Frame, D., & Pierrehumbert, R. (2021). Agriculture's contribution to climate change and role in mitigation is distinct from predominantly fossil CO2-emitting sectors. *Frontiers in Sustainable Food Systems, 1–9*. https://doi.org/10.3389/fsufs.2020.518039

Maravegias, N., & Doukas, Y. E. (2011). Traceability and the new CAP. In G. Baourakis, K. Mattas, C. Zopounidis, & G. van Dijk (Eds.), *A resilient european food industry in a challenging world. (European political, economic, and security issues; agriculture issues and policies)* (pp. 245–253). Nova Science Publishers. ISNB: 9781611220322.

Maravegias, N., Doukas, Y. E., & Petides, P. (2023). The political economy of the common agricultural policy's green architecture. *Sustainable Development, Culture, Traditions Journal, 1*(B), 73–84. https://doi.org/10.26341/issn.2241-4010-2023-1b-

OECD. (2019). *Trends and drivers of agri-environmental performance in OECD countries*. OECD Publishing. https://doi.org/10.1787/b59b1142-en

OECD. (2022). *Agricultural Outlook 2022–31*, from https://www.oecd-ilibrary.org/agriculture-and-food/oecd-fao-agricultural-outlook-2022-2031_f1b0b29c-en

Pe'er, G., Bonn, A., Bruelheide, H., et al. (2020). Action needed for the E.U. Common Agricultural Policy to address sustainability challenges. *People and Nature, 2*(2), 305–316. https://doi.org/10.1002/pan3.10080

Powell, J. P., & Reinhard, S. (2016). Measuring the effects of extreme weather events on yields. *Weather and Climate Extremes, 12*, 69–79. https://doi.org/10.1016/j.wace.2016.02.003

Roberts, T. J., et al. (2021). Rebooting a failed promise of climate finance. *Nature Climate Change, 11*(3), 180–182. https://doi.org/10.1038/s41558-021-00990-2

Schimmelpfennin, G. D., & Thirtle, C. (1999). The Internationalization of Agricultural Technology: Patents R&D Spillovers, and their effects on productivity in the European Union and United States. *Contemporary Economic Policy, 17*(4), 457–468.

Siardos, G., & Koutsouris, A. (2004). *Αειφορική γεωργία και ανάπτυξη* (2nd ed.). Zigos. ISBN: 9608065348.

United Nations. (2021). *Goal 13: Take urgent action to combat climate change and its impacts*. Department of Economic and Social Affairs. United Nations. https://sdgs.un.org/goals/goal13

Verschuuren, J. (2022). Achieving agricultural greenhouse gas emission reductions in the E.U. post-2030: What options do we have? *RECIEL, 31*(2), 246–257. https://doi.org/10.1111/reel.12448

World Economic Forum. (2022). *What did COP27 accomplish and what actions can we expect as a result?* https://www.weforum.org/agenda/2022/11/what-happened-cop27-climate-change-what-is-next/

Wreford, A., Moran, D., & Adger, N. (2010). *Climate change and agriculture: Impacts, adaptation and mitigation*. OECD Publishing. https://doi.org/10.1787/9789264086876-en

Building Climate-Resilient Food Systems: The Case of IFAD in Brazil's Semiarid

Alexandra Teixeira and Camila Amorim Jardim

Introduction

The world is not in the right direction in the global call to end hunger, achieve food security and improved nutrition, and promote sustainable agriculture (SDGs 1 and 2) by 2030. Climate change disproportionately affects the most vulnerable rural communities and is one of the most significant obstacles in this path. Agriculture has been subjected to increased extreme events, such as extended periods of drought and erratic temperatures. Those events harm agricultural production systems; drive land use changes; damage infrastructure; boost the risks of pests and diseases; disrupt pollination, flowering, and fruiting processes; and increase soil erosion and degradation (Lengnick, 2022). The combined negative impacts of climate change in food systems (FSs) further exacerbate rural populations' poverty and food and nutrition insecurity.

A. Teixeira (✉)
International Fund for Agricultural Development, Brazil, Rio de Janeiro

C. A. Jardim
International Relations Institute, University of Brasília, Brasília, Brazil

© The Author(s) 2024
A. Ribeiro Hoffmann et al. (eds.), *Climate Change in Regional Perspective*,
United Nations University Series on Regionalism 27,
https://doi.org/10.1007/978-3-031-49329-4_10

In this context, Food Nutrition Security (FNS) is an integrated approach combining two underlying concepts: food security and nutrition security.[1] While food production and consumption are key drivers of climate change, undernutrition undermines climate resilience – the extent to which social or ecological systems can maintain, recover, and improve their integrity and functionality when subject to disturbance[2] – and the coping strategies of vulnerable populations. In this context, IFAD has set the target that at least 50% of all new projects should be nutrition-sensitive[3] (IFAD, 2019).

Resilient food systems (FSs) lie at the heart of the nutrition-climate nexus. The food we eat, how it is produced, and the journey from farm to plate determine how FSs affect human and planetary health. Climate variability and extremes are key drivers behind the worsening of global food insecurity and malnutrition, which are also exacerbated by other crises, such as the COVID-19 pandemic, the economic slowdown, and the inflationary impacts of the conflict in Ukraine.[4]

Latin America and the Caribbean (LAC), although expected to account for 25% of global agricultural and fisheries exports by 2028 (OECD/FAO, 2019), has the highest cost of an adequate diet in the world (FAO et al., 2022). In the region, Brazil is the largest food exporter and, yet, had over 33 million people suffering from hunger in 2022 (Vigna, 2022). Therefore, LAC is no exception to the greater world tendency: hunger, food insecurity, child overweight, and adult obesity are all worsening. (FAO et al., 2023).

Regionalism can either contribute to the worsening of this scenario or act positively in fighting it. While the liberalization of commerce can, on the one hand, provide cheaper and better products to the final costumer, it can also deepen the social and economic asymmetries between developed and developing countries, in a context which the first offer manufactured products, while the second provide commodities and suffers with the deterioration of terms of trade. The higher demand for commodities in developing countries can increase the pressure for deforestation

---

[1] The concept of food security evolved from "freedom from hunger" (SEN, 1987) into a broad concept, achieved when individuals have access to sufficient and nutritious food in adequate quantities. Nowadays, it encompasses four dimensions: (i) availability, (ii) access, (iii) utilization, and (iv) stability. Nutrition security, in its turn, evolved from the multisectoral nutrition planning approach and UNICEF's conceptual framework, now implying constant and equitable access to healthy, safe, sustainable, and affordable food that is essential for a healthy and high-quality life, therefore assessing the nutritional quality of food intake. It has three determinants: (i) access to adequate food, (ii) care and feeding practices, and (iii) sanitation and health.

[2] "Resilience" has three main pillars: maintaining functionality, improving, and transitioning to a better-off state (IFAD, 2015).

[3] A nutrition-sensitive project addresses the underlying causes of malnutrition related to inadequate household food security, maternal and child care, and environmental health (IFAD, 2018). According to Njoro (2021), "Nutrition-sensitive agriculture is a food-based approach to agricultural development that puts nutritionally rich foods and dietary diversity at the heart of overcoming undernutrition, overnutrition, and micronutrient deficiencies."

[4] The rise in international food prices and the effects of food inflation have increased the costs and the unaffordability of a healthy diet worldwide, and LAC is the worst hit. The region has the highest cost of a healthy diet compared to the rest of the world today (FAO et al., 2023).

and climate change in their territories, reinforcing the very problematic trend of violence related to land, biodiversity loss, desertification, change in rain patterns, worse working conditions, pollution, etc. In a national – and regional – economic perspective, this pattern also results in macroeconomic vulnerability to external shocks and the oscillations on commodities prices.

In this context, regional cooperation between LAC countries can provide policy coordination and investment in the diversification of the production patters, developing technologies and value chains with higher value-added but that, first and foremost, is compatible with a sustainable future. On the other hand, the cooperation between CELAC countries and the European Union has the potential to promote commercial trends that support sustainable development, sharing technology and financing the green transition in those countries. This movement should be cautious guaranteeing that European countries are not using LAC to export their harmful modes of production, such as chemicals, pollution, or the pursue for cheaper resources, including land and labor, an inherently neocolonial pattern that can be reproduced through regionalism and economic cooperation.

Given this broader scenario, this chapter proposes the right to adequate food, guided by a human rights-based approach (HRBA) to development (Cornwall & Nyamu-Musembi, 2004),[5] to be placed at the center of strategies to mitigate and adapt to the impacts of climate change in regional contexts. It also advocates for a transformative adaptation of agriculture in response to the current effects of climate change, analyzing transformative pathways and possible solutions for building climate-resilient FSs in LAC.

Given the imperative of constructing climate-resilient FSs, considering its key challenges and best practices, this chapter presents some experiences of the International Fund for Agricultural Development (IFAD),[6] both a specialized agency of the United Nations (UN) and an International Financial Institution (IFI). IFAD seeks to transform rural economies and FSs by making them more inclusive, productive, resilient, and sustainable, investing in the millions of rural people who are most at risk of being left behind: in poverty, small-scale food producers, women, youth, persons with disabilities, and other vulnerable groups living in rural areas (home of ¾ of the world's population in poverty) (UNDESA, 2021). IFAD is the only specialized global development organization exclusively focused on and dedicated to transforming agriculture, rural economies, and FSs, reaching the remotest rural areas (IFAD, 2023). It advocates for a comprehensive and participatory

---

[5] The right to adequate food is realized when every man, woman, and child, alone or in a community with others, has physical and economic access at all times to adequate food or means for its procurement. This right is recognized in the 1948 Universal Declaration of Human Rights as part of the right to an adequate standard of living. It is also enshrined in the 1966 International Covenant on Economic, Social, and Cultural Rights. Regional treaties and national constitutions also protect it (OHCR, 2010).

[6] IFAD's creation was one of the major outcomes of the 1974 World Food Conference, organized in response to the global food crises of the early 1970s, when food shortages were causing widespread famine and malnutrition, especially in Africa's Sahel.

approach to strengthening food and nutrition security and, therefore, targets structural causes, such as socioenvironmental conditions, access to drinking water, and breastfeeding.

The chapter emphasizes the imperative to accelerate and scale up actions for transforming food systems (FSs)[7] toward resilience in response to climate change. It also highlights pathways to build climate resilience based on existing global policy platforms and solutions and best practices being implemented in LAC. To illustrate those points, we discuss IFAD's Pro-Semiarid Project in Bahia (PSA), Brazil, the organization's best-ranked project in LAC and the world's second-best. The chapter is divided into six parts. Beyond this introduction, section two discusses the nutrition-climate change nexus, presenting the key concepts of food security (FS) and climate resilience. Section three debates the role of international organizations and the urge to increase synergies and reduce organizational silos among different institutions and initiatives. Section four presents key pathways to building climate-resilient FSs, considering global and local spheres. Section five takes the key points discussed in the previous ones to analyze PSA. Finally, section six presents a conclusion with the main topics raised throughout the text, mainly the centrality of the nutrition-climate change nexus and the impossibility of dealing with one without properly also targeting the other.

The Nutrition-Climate Change Nexus: Defining Sustainable Food Systems and Climate Resilience

Over halfway through the 2030 Agenda for Sustainable Development term, the global community is still far from reaching its 17 goals (SDGs). As a pathway to achieve them, nutrition security and climate change are now seen as deeply connected, as climate change is both a result of the existing FSs and, on the other way around, climate change outcomes also drive change in FSs (Bakker et al., 2021).

According to IPCC (2021), "climate resilience" refers to the capacity to avoid poverty in the face of climate-related shocks or climate extremes, often referred to as extreme weather, extreme weather events, or extreme climate events. Another related concept is "climate stresses," understood as persistent occurrences of lower-intensity climate hazards (i.e., low-intensity/high-frequency damaging phenomena), such as soil erosion, salinization of soils, and groundwater, a shift of river runoff patterns, migration of species, or a rise in sea level (IFAD, 2015).

Food systems consist of a set of interlinked actors of food products that offer value-added activities in "the production, aggregation, processing, distribution, consumption, and disposal (loss or waste)" (Von Braun et al., 2021), coming from

---

[7] Food systems cover all the components needed from producing to consuming foods and the management of waste and by-products, and typologies include modern, mixed, and traditional food systems (HLPE, 2017).

"agriculture (including livestock), forestry, fisheries, and food industries, and the broader economic, societal, and natural environments in which they are embedded" (Von Braun et al., 2021). They exist at different levels, from local to global, including their actors' values and cultures. FSs can change – and are expected to change, both for planned (i.e., 2030 agenda) and unplanned reasons (climate shocks) – and change can come either from external causes (e.g., conflicts) or internal ones (i.e., increased productivity due to innovations).

Along the same lines, sustainable FSs are the ones that contribute both to food and nutrition security "in such a way that economic, social, cultural, and environmental bases to generate FNS for future generations are safeguarded" (Von Braun et al., 2021). However, sustainable FSs do not necessarily guarantee good nutrition, as contexts of sanitation, infectious diseases, hygiene, access to clean drinking water, adequate child care, and access to nutritious food are also essential. Furthermore, the definition of sustainable FSs is not fixed, as it reflects a relative change in comparison with a previous scenario and, thus, presents a horizon and parameters to guide political action (Von Braun et al., 2021).

Climate change directly affects all forms of malnutrition, particularly undernutrition, not only because it reduces food availability in low- and middle-income countries (LMICs) but also because increased CO_2 levels reduce iron, zinc, and protein levels in staple crops, decreasing the nutritional value of crops such as wheat, rice, potatoes, soy, and peas. In contrast, nutrient-rich foods are frequently susceptible to water constraints. Furthermore, higher temperatures might cause a reduction in the soil's decomposition of organic matter, which results in lower fertility and reduced water-retaining capacity of the soil, aggravating desertification processes (Bakker et al., 2021: 22–23; Soares et al., 2019).

Another aggravating factor is the restricted variety of food that our FSs rely on. Approximately 75% of the planet's food production is focused on 12 plants and 5 animal species, making our FSs highly susceptible to supply shocks. Those shocks can have multiple factors, including extreme weather and climate-related events, such as heat waves, drought, floods, and strong winds, or even the climate-related spread of pests and diseases into new geographical regions (Bakker et al. 2021).

According to Bakker et al. (2021: 10) "food production and consumption have major impacts on environment-related sustainable development goals (SDGs 6, 7, 9, 12, 13, 14, and 15)." Humanity has reached increased food production, mostly due to agricultural intensification practices and innovations. However, our population is growing fast, with an expected increase of 33% in 30 years (Soares et al., 2019: 2). By 2050, humanity will need an increase of 60% in agricultural production, considering that the world's population might have reached 9.7 billion (Ruiz et al., 2020).

The increase in agricultural productivity in the past decades did not reflect greater nutrition patterns. Access to sufficient, safe, and nutritious food that meets dietary needs and preferences is conditioned to the price and affordability of diets. Considering different types of diets (energy-sufficient, nutrient-adequate, and healthy diets), the global average cost that meets daily energy needs is estimated to be USD 0.79 per day, whereas meeting all essential nutrient requirements is approximately USD 2.33, and healthy diet median global costs of USD 3.75 a day, much higher than the poverty line of USD 1.90. Thus, healthy diets are much more expensive than the daily food expenditures of most people in the Global South. In LAC, healthy diets are unaffordable for over 20% of the population, and education and individual behavior change are insufficient to solve nutrition security issues. To make sure its goals are achieved, prices should drop, and increased local production and harvesting should be achieved (Herforth et al., 2020: xi-xii).

Malnutrition also causes deep social and economic costs. Child undernutrition is responsible for some African countries to lose up to one-sixth of their annual GDP. Child stunting compromises both physical and cognitive capabilities, and undernutrition is known to reduce a nation's economic growth by at least 8% due to cognitive, productive, or reduced schooling losses. FNS issues also increase the costs of nutrition-related illnesses, raising comorbidities, and multiple forms of climate-related and nutritional health risks, considering both communicable (parasitic, viral, and bacterial diseases) and noncommunicable diseases (diabetes, cardiovascular, respiratory, etc.) (Bakker et al., 2021: 11).

Global amounts of government's financing of food and agriculture have reached nearly USD 630 billion per year. Nonetheless, those investments are not being adequately directed to sustainable and resilient agri-food systems, as a large amount of those is responsible for distorting market prices, jeopardizing small-scale producers and Indigenous Peoples, and yet not delivering healthy diets. Cereals have been highly subsidized by food-importing countries, favoring the production of those and making pulses, seeds, fruits, and vegetables less profitable. Policies like this have improved calorie intake but have disfavored improved nutrition and health outcomes, mainly among vulnerable populations (FAO et al., 2022).

Meeting the growing demand for food supplies while eradicating hunger and undernutrition might be one of the humanities' greatest challenges in the twenty-first century, and those goals will definitely not be achieved if our policies do not keep an approach that takes climate change, sustainability, and climate resilience into consideration. Agriculture is responsible for about 20% of human greenhouse gas emissions (GHGs), with meat and dairy products being the ones with the highest carbon footprint; however, those are responsible for a great amount of vulnerable populations' nutrient intake. Land use change, mostly driven by agriculture, is responsible for 15 to 17% of emissions. Furthermore, not only CO_2 emissions (from deforestation, food processing, transportation, etc.) are worrisome. Flooded rice fields and livestock are highly responsible for methane (CH_4) emissions, while organic and inorganic nitrogen fertilizers release nitrous oxide (N_2O) into the atmosphere. Beyond GHGs, food production also relies on excessive use of water and farmland and is responsible for biodiversity loss, having a very large amount of systemic impacts (Bakker et al., 2021: 22–23).

Hence, there has been increasing interest in identifying climate change mitigation and adaptation measures offering nutrition co-benefits (and vice versa), seeking transformation towards climate-smart[8] and nutrition-sensitive[9] FSs, considering that climate change increases hunger, undernutrition, and poverty. For this, studies have been looking at different areas of the FSs, including matters such as the food supply chain, food environment,[10] and consumer behavior and diets (Bakker et al., 2021). Therefore, healthier diets can be seen as FSs outcomes, as nutrition and dietary patterns are determined by and determining FSs.

The Role of International Organizations: Fostering Synergies for Nutrition Security in Latin America and the Caribbean

Among the greatest challenges to achieving SDGs 1 and 2, there is the lack of integrated approaches and difficulty in achieving greater synergies between the various projects being implemented by diverse groups and institutions. The UN Rome-based agencies (RBAs), Food and Agriculture Organization of the United Nations (FAO), IFAD, and the World Food Programme (WFP), are central to the organization's development, humanitarian and resilience assistance to the thematic areas of food, agriculture, and transformative rural development. For achieving SDG 2, enhanced synergies among the RBAs are crucial, as they share the common vision of ending hunger, malnutrition, and promoting sustainable agriculture and rural transformation (WFP, 2023).

The agenda is urgent. In 2021, 3.1 billion people could not afford a healthy diet[11] (including 80% of the population in Africa). Globally, stunting among children under 5 years old decreased from 26.2% in 2012 to 22% in 2020; however, Africa remains the highest at 30.9%, followed by Asia at 21.2% (FAO et al., 2022). An estimated 30% of the world's population faces micronutrient deficiency, and some 676 million are obese. Projections are that nearly 670 million people will be facing hunger in 2030. Several factors have contributed to this situation, including quality of diets; gender inequality; food availability, affordability, and accessibility; global nutrition financing; and climate change. Regarding climate change and its increasing impacts on FNS, approximately 80% of global cropland and 60% of global food

---

[8] According to FAO's website, "climate-smart agriculture (CSA) is an approach that helps guide actions to transform agri-food systems towards green and climate resilient practices." Available at https://www.fao.org/climate-smart-agriculture/en/

[9] To reach a nutrition-sensitive food system, national policies and investments should be reviewed to integrate nutrition objectives into food and agriculture policy, program design, and implementation (FAO, 2015).

[10] The concept of food environment "refers to the physical, economic, political, and socio-cultural surroundings, opportunities and conditions that create everyday prompts, shaping people's dietary preferences and choices, as well as nutritional status" (HLPE, 2017: 28).

[11] A healthy diet consists of adequate calories, essential nutrients, and diverse foods from several food groups needed for an active and healthy life. According to the WHO and FAO, healthy diets consist of a wide variety of unprocessed or minimally processed foods and are balanced among all food groups, including a minimum of five servings of fruits and vegetables per day.

output is a result of rainfed agricultural production, which can be dramatically affected by changes in water availability or transformations of the water cycle and rainfall patterns (FAO, 2017).

Regionally, LAC faces considerable challenges in eradicating hunger and malnutrition in all its forms. Despite the progress made in the region to reduce child undernutrition in the past decades, hunger and food insecurity have risen since 2014, reaching their highest levels during the COVID-19 pandemic. The increase in the proportion of people experiencing hunger during the pandemic was more significant in the region than at the global level. Between 2019 and 2021, the regional prevalence of hunger increased by 28%, compared to a global increase of 23% (FAO et al., 2023). In 2021, food insecurity affected 40% of the people in LAC (about 267 million people), compared to a global prevalence of 29.3%. Currently, the region has the highest cost of an adequate diet compared to the rest of the world,[12] increasing the vulnerability to malnutrition in all its forms and to noncommunicable diseases (NCD).

This situation disproportionately affects those living in the most vulnerable situations being particularly risky for rural people, women, children (especially girls), Indigenous Peoples, Afro-descendants, and traditional communities. Climate change can strongly increase the already existing heavy workload of women, which generates negative impacts on child care and raises the risk of undernutrition (Bakker et al., 2021: 12). There is also a clear gender gap in food insecurity in the world, with an aggravation tendency since the COVID-19 pandemic. In 2021, while 27.6% of men in the world had some degree of food insecurity, for women, this number reached 31.9%, 1% higher than the previous year (FAO et al., 2022: xvii).

In LAC, the nutrition crisis worsens, while adverse weather conditions cause harvests to fail, resulting in hunger, malnutrition, and loss of livelihoods (Lengnick, 2022). Stress to water systems and sanitation threatens the quantity and quality of water available for irrigation and human use. Additional climate change impacts in FSs are yet to come. In LAC, the average temperature is increasing and will continue to grow faster than the global average (IPCC, 2021), which makes it imperative that different sectors are involved in conceiving, testing, implementing, and improving climate change mitigation and adaptation measures with nutrition co-benefits. The current regional FSs are associated with significant environmental externalities as agricultural activities drive land use changes that cause biodiversity loss and environmental degradation. In this context, deep transformations in FSs are required to ensure sustainability.

It is important to highlight that if humanity does not develop resilient FSs, it will be increasingly difficult to produce food in adequate amounts and quality in the following decades. In that regard, several solutions are being proposed, including conservation farming practices to enhance soil organic carbon, diversified agroecological backyards, biofortification, and Agroforestry Systems, which generate multiple

---

[12] This indicator, calculated by FAO, identifies the least-cost healthy diet available at each given time and place that meets food-based dietary guidelines (FBDGs) recommendations (FAO et al., 2022).

benefits for nutrition, food security, and the environment through increasing crop productivity and land carbon sinks (Frank et al., 2017).

There are critical steps to make sure the progress that has already been made in reducing all forms of malnutrition is not lost, such as (i) breaking down organizational silos (separated initiatives of action) for an increased synergy among different actors in initiatives on food and agriculture, urban design, and land use; (ii) increased data collection and analysis, including monitoring and evaluation (M&E) improvement; (iii) increased financing for climate-resilient FS and nutrition from diverse sectors, including private partners; (iv) focusing on healthy diets to improve nutrition all over the world; and (v) developing better and more ambitious targets and goals (Bakker et al., 2021). This discussion will be further developed in the following sections.

Key Pathways for Climate-Resilient Food Systems

In 2021, the Food Systems Summit emphasized the need for systems-level change. Seeking to raise awareness on how transforming our FSs can support the achievement of the SDGs, global leaders backed promoting holistic and inclusive food systems-based approaches to poverty alleviation, nutrition, resilient and reliable agricultural production, resource conservation, and climate change mitigation and adaptation. The Summit established five Action Tracks (AT)[13] intended to highlight essential pathways to transform FSs and reach the 2030 Agenda objectives.[14] Together, the ATs explore how key levers of change – human rights, innovation, finance, gender equality, and women's empowerment – can be mobilized to meet the Summit's objectives. Each lever of change can bring about significant progress on both FSs transformation and achieving all 17 Sustainable Development Goals (SDGs).

In LAC, regional agreements have been set regarding the challenges and opportunities for building more inclusive, sustainable, and resilient FSs to support the goals set in the United Nations Food Systems Summit 2021. Sixteen LAC countries have joined 12 different coalitions, and 14 countries presented roadmaps at the Summit.

Taking inspiration from the Food System Summits' levers of change, we propose five key points of regional action to promote more climate-resilient FSs:

(i) *Multi-stakeholder partnerships and intersectoral approach:* Multisectoral alliances with academia, public, private, and third-sector institutions can

---

[13] The Action Tracks are as follows: (1) ensure access to safe and nutritious food for all; (2) shift to sustainable consumption patterns; (3) boost nature-positive production; (4) advance equitable livelihoods; and (5) build resilience to vulnerabilities, shocks, and stress.

[14] Over 2000 ideas were received by the Action Tracks and have now been consolidated into 15 Action Areas. Source: UN Food Systems Summit 2021. Available at: Action Tracks | United Nations

strengthen the project's design, implementation, and M&E, thus, maximizing results on both climate resilience and food security and nutrition, as well as development effectiveness.

(ii) *Nutrition-sensitive and climate-resilient value chains:* The value chain approach has traditionally focused on increasing economic returns for producers, even when working with small-holder farmers, but they also offer opportunities to ensure diverse, nutritious, and safe foods are accessible to everyone. Thus, it is necessary to repurpose resources to prioritize food consumers and incentivize sustainable production, supply, and consumption of nutritious foods to make healthy diets more affordable. Interventions prioritizing the adaptation needs of small-scale producers and micro-, small-, and medium-sized enterprises (MSMEs) along food supply chains can also help to ensure the affordability of healthy diets while bolstering the resilience and inclusiveness of agri-food systems. Innovative governance mechanisms give a real voice and influence to poor rural people, including small-scale producers.

(iii) *Finance gaps:* There is a global need to close the financial gap for climate-resilient food systems. Today, an extra USD 39–50 billion is annually needed to meet both nutrition-specific and nutrition-sensitive needs until 2030 (Global Nutrition Report, 2021). Small-scale producers remain underserved by global climate finance. They bear the devastating consequences of changing climate, degraded soils, food insecurity, and irregular migration. So far, only about 1.7% of the money invested globally in climate finance reaches small-scale producers, and it mostly goes to mitigation objectives compared to adaptation. Possible solutions for this could be as follows:

- Proactively increasing the number of climate finance projects integrating nutrition and improving monitoring and evaluation (M&E).
- Exploring opportunities to mobilize foundations and private sector resources to help frame commitments from the private sector to support small to medium enterprises (SMEs).
- Pursuing, in partnership with governments, the feasibility of nutrition-focused development impact bonds to deliver outcomes for nutrition.

(iv) *Gender equality and women's empowerment (GEWE):* Women, during their reproductive age (15–49 years), bear a disproportionate burden of malnutrition, and they are also often the most vulnerable to climate change impacts (Tantoh et al., 2022). Lessons emerging from UN agencies indicate that improving individual women and girls' nutrition without addressing the discriminatory gender norms and unequal power imbalances that contribute to gender inequality and malnutrition is insufficient and many times more harmful (FAO et al., 2022).

(v) *Social behavior change communication (SBCC)* is a successful and strategic regional approach being scaled up by multiple institutions, including IFAD. SBCC interventions include increasing consumer demand for healthy diets, particularly in rural areas, through information, education, and awareness. Behavior change is needed all along the food chain. Achieving necessary behavioral changes to improve nutrition is only possible with a dedicated theory of change that covers many dimensions, such as education, consumer awareness, and market incentives to facilitate nutritional change. Among the existing examples, there are successful programs on school health and nutrition to increase consumer education and outreach efforts, such as PNAE in Brazil. SBCC is also fundamental for changing knowledge, behavior, and practices regarding climate change. This approach could emphasize the climate-nutrition nexus.

Case Study: IFAD's Pro-Semiarid Project in Bahia, Brazil

Recent projects coordinated by IFAD have presented key insights into promoting FSs that are resilient, sustainable, healthy, and nutrition-sensitive. Among those, it is worth mentioning the following: (a) strategic alliances and multisector cooperation to improve climate resilience and nutrition; (b) adequate targeting of populations in greatest vulnerability; (c) integration of resilient agricultural practices with social infrastructure; (d) rescue of traditional knowledge and species to increase dietary diversity; and (e) leadership, autonomy, and empowerment of women and girls. To illustrate those understandings and deepen their discussion, this section presents the case study of the Rural Sustainable Development Project in the Semiarid Region of Bahia (Pro-Semiarid – PSA).[15]

The PSA aims to reduce poverty in rural areas of the Brazilian state of Bahia by increasing production, creating agricultural and nonagricultural work opportunities, and developing human and social capital. It seeks to strengthen the capacities of individuals, communities, and economic organizations in rural areas to support the development of sustainable productive activities and their insertion into value chains and markets. It operates in 32 municipalities with high incidences of poverty and vulnerability in the semiarid area of northern Bahia State, giving priority to women, youth, indigenous peoples, and traditional communities.

PSA interventions were centered on building resilient FSs, having strong co-benefits for climate and nutrition. It is currently the second most well-rated IFAD project in the world and the first best evaluated in LAC because it was able to foster the food system's resilience using a bottom-up, community-based, nutrition-sensitive, and gender-inclusive approach, implementing actions to reduce the community's vulnerability to the impacts of climate change while contributing to building resilient livelihood systems and FNS co-benefits.

PSA's subprojects are organized in rural territories composed of contiguous or nearby communities represented by an organization, usually constituted of four communities. The territorial investment plans were built on a diagnosis of the territory's environmental, social, and productive needs, strengths, and weaknesses.

---

[15] For more details on the project, access: https://www.ifad.org/en/web/operations/-/project/1100001674

These territorial plans acted as master plans to guide the project's collective and individual investments. This very effective methodology ensured participation, precise targeting, and demand-driven investments and that territories with a large proportion of rural people in poverty and food insecurity were systematically reached.

The project's targeting strategy allowed for serving the communities most vulnerable both to climate change and malnutrition. Regarding food security, from 2016 to 2021, the project's interventions generated important positive results: the percentage of families in a food security situation increased by 10.95% (from 69.35% to 76.96%), severe food insecurity was reduced by 84.21%, and moderate insecurity was reduced by 31.89%. Regarding nutrition, the impact evaluation (IE) indicated that a higher proportion of families in the treatment group (TG) consume food from agroecological gardens compared to the control group (CG), suggesting a greater intake of diversified, safe, healthy, and nutritious food. In the TG, there was also an 11% increase in the number of families that always have a diversified diet, compared to a 4% decrease in the CG between 2016 and 2021.

The agroecological perspective guiding PSA ensures communities adapt to and mitigate climate change and also guarantees sustainable FSs that produce nutritious, healthy, diversified, and safe food, improving the nutritional benefits of the partner population. Such an approach aims to build local communities' capacities, knowledge, and social capital and jointly develop solutions appropriate to the local context and reflect the interests and objectives of project participants. In addition, the significant investment in capacity-building of agricultural technicians and farmers in agroecological practices through the agroecology and coexistence with the semiarid region study groups (NEACS) and the use of tools and methodologies (Agroecosystems Sustainability Indicators, ISA; Agroecological Transition Indicators; and Method of Economic-Ecological Analysis of Agro-Ecosystems, LUME) allowed the Project Management Unit (PMU) to monitor progress and identify gaps in both project activities and agro-ecosystems in the transition to agroecological and resilient FSs.

Although PSA supported building climate-resilient FSs through its agroecological approach and its lessons are taken into consideration in the design, implementation, monitoring, and evaluation (M&E) of new IFAD projects, there are still key challenges. At design, PSA was not classified as nutrition-sensitive nor climate finance, as it did not include particular strategies for both themes or any specific nutrition and climate indicators.[16] The Project Management Unit (PMU) did not have a specialist in nutrition.

The selected case study offers a glimpse of how development actors are trying to improve nutrition co-benefits as they work to increase FSs' climate resilience. In addition to climate and nutrition synergies, the case identifies synergies with gender and environmental sustainability goals. As discussed, smallholder farming practices are becoming increasingly challenging in the context of changing climate, often making woman's already heavy workload even heavier, which can further increase women's needs for a nutritious diet, which may not always be easy in many cultural

---

[16] IFAD has two core indicators (COI) related to nutrition, the MDD-W (minimum dietary diversity for women) and the KAP (knowledge, attitude, and practices), in terms of climate change.

contexts. This can also reduce women's time for caring activities, including preparing healthy meals. The nexus between climate change, nutrition, and gender is, therefore, a key one. Also, evidence- and knowledge-based project design, implementation, and M&E approaches are also crucial to delivering multiple co-benefits, which have been observed in PSA so far.

Conclusion

Nutrition has been increasingly seen as a structuring issue of the 2030 Agenda, not only concerning hunger or nutrition insecurity eradication but also regarding nations' economic development and climate change mitigation and adaptation. Climate change is a result of the existing FSs, as agriculture is responsible for 20% of GHGs, but it also impacts FSs – considering that it reduces food availability and its nutritional value. This perception is being reflected in UN agencies focused on food, agriculture, and transformative rural development, such as IFAD.

Regionalism and the trade pattern it promotes can either contribute to fight this scenario or to its worsening. In this context, the concept of food systems (FSs) is key. It encompasses all actors and their interlinked value-adding activities involved in the food chains, including production, aggregation, processing, distribution, consumption, and disposal of food products. Enhanced synergies among the various actors involved in rural development are needed, and the reduction of organizational silos at the regional level is essential to make sure that the most vulnerable are properly targeted and that transformation toward more climate-resilient FSs is possible. In a macro-perspective, it is important that economic and commercial relationships among LAC countries and the European Union does not follow a neocolonial pattern, in which the developing countries are trapped in offering commodities, pushing their resources to the limit under a strictly extractive perspective, while importing technological products, and, in a context of climate change, also pollution, carbon, and other harmful substances that are not allowed anymore in European countries.

Although humanity has achieved increased agricultural productivity, healthy diets are still unaffordable to a great part of the Global South and to 20% of the population in LAC. Education and individual behavior change must be added up with active measures to reduce food prices, including sustainable and climate-resilient local production and financing. IFAD's project in Bahia (Brazil) illustrates possible pathways, showing the relevance of a holistic approach that empowers vulnerable populations, increasing their capacity for production, their awareness of issues related to climate change and nutrition security, and their digital inclusion and improved commercial partnerships.

Therefore, the right to adequate food needs to be placed at the center of strategies to mitigate and adapt to the impacts of climate change and fight against hunger and poverty. The nutrition-climate nexus increasingly structures global climate action. Thus, nutrition-sensitive climate finance and improved cross-sectoral collaboration are proposed means to address climate change impacts and FNS.

References

Bakker, S., Macheka, L., Eunice, L., Koopmanschap, E., Bosch, D., Hennemann, I., & Roosendaal, L. (2021). *Food-system interventions with climate change and nutrition co-benefits: A literature review.* Available at: https://library.wur.nl/WebQuery/wurpubs/fulltext/547743

Cornwall, A., & Nyamu-Musembi, C. (2004). Putting the 'rights-based approach' to development into perspective. *Third World Quarterly, 25*(8), 1415–1437.

FAO. (2015). *Designing nutrition-sensitive agriculture investments: Checklist and guidance for programme formulation.* Food and Agriculture Organization of the United Nations. Available at: https://www.fao.org/3/i5107e/i5107e.pdf

FAO. (2017). *Water for sustainable food and agriculture: A report produced for the G20 presidency of Germany.* Food and Agriculture Organization of the United Nations. Available at: https://www.fao.org/3/i7959e/i7959e.pdf

FAO, IFAD, PAHO, UNICEF, and WFP. (2023). *Regional overview of food security and nutrition – Latin America and the Caribbean 2022*: Towards improving the affordability of healthy diets. Santiago. https://doi.org/10.4060/cc3859en.

FAO, IFAD, UNICEF, WFP and WHO. (2022). The state of food security and nutrition in the world 2022. Repurposing food and agricultural policies to make healthy diets more affordable. Rome, FAO. https://doi.org/10.4060/cc0639en.

Frank, S., Havlík, P., Soussana, J.-F., Levesque, A., Valin, H., Wollenberg, E., Kleinwechter, U., Fricko, O., et al. (2017). Reducing greenhouse gas emissions in agriculture without compromising food security? *Environmental Research Letters, 2017, 12*(10). https://doi.org/10.1088/1748-9326/aa8c83

Global Nutrition Report. (2021). Global nutrition report: The state of global nutrition. Bristol, UK: Development Initiatives. Available at: https://globalnutritionreport.org/reports/2021-global-nutrition-report.

Herforth, A., Bai, Y., Venkat, A., Mahrt, K., Ebel, A. & Masters, W.A. (2020). Cost and affordability of healthy diets across and within countries. Background paper for the state of food security and nutrition in the world 2020. *FAO agricultural development economics technical study* No. 9. Rome, FAO. https://doi.org/10.4060/cb2431e.

HLPE. (2017). *Nutrition and food systems.* A report by the high-level panel of experts on food security and nutrition of committee on world food security, Rome. Available at: https://www.fao.org/3/i7846e/i7846e.pdf.

IFAD. (2015). *How to do: Measuring climate resilience.* Environmental and climate change (Rome: IFAD, 2015). Available at: www.ifad.org/documents/38714170/40193941/htdn_climate_resilience.pdf/fd0b42b0-3fc1-41e2-bd45-c66506fa5004

IFAD. (2018). *Nutrition-sensitive value chains*: A guide for project design. Volume I. Rome, 2018. https://www.ifad.org/documents/38714170/40804965/GFPD+Nutrition-sensitive+value+chains+VOL.1.pdf/5177a3c0-a148-4b1f-8fff-967a42f51ce8?t=1584027322000

IFAD. (2019). *IFAD Action Plan Nutrition: 2019–2025.* Available at: https://www.ifad.org/en/-/document/ifad-actionplan-nutrition-2019-2025.

IFAD. (2023). *Vision: Rural transformation that lasts.* Available at: https://www.ifad.org/en/vision.

IPCC. (2021). *Climate change 2021: The physical science basis.* Contribution of working group I to the sixth assessment report of the intergovernmental panel on climate change, Cambridge University Press.

Lengnick, Laura. (2022). *Resilient agriculture: Cultivating food systems for a changing climate* (2nd expanded edition). Gabriola Island: New society.

Njoro, Joyce. (2021). *Nutrition-sensitive agriculture: The cornerstone of a healthier world.* December 7, 2021. https://www.ifad.org/en/web/latest/-/blogs/nutrition-sensitive-agriculture-the-cornerstone-of-a-healthier-world

OECD/FAO. (2019). *OECD-FAO Agricultural Outlook 2019-2028.* OECD Publishing. https://doi.org/10.1787/agr_outlook-2019-en

OHCR. (2010).*The right to adequate food. United Nations human rights.* Office of the High Commissioner for human rights. Fact sheet no. 34. Geneva. Available at: FactSheet34en.pdf (ohchr.org).

Ruiz, N., Noe-Bustamante, L., & Saber, N. (2020. Coming of age. IMF. Available at: https://www.imf.org/en/Publications/fandd/issues/2020/03/infographic-global-population-trends-picture#:~:text=In%20just%2030%20years%2C%20the,to%20reach%20just%209.7%20billion.

Sen, A. (1987). *Food and freedom.* Sir John Crawford Memorial Lecture.

Soares, J. C., Santos, C. S., Carvalho, S. M. P., et al. (2019). Preserving the nutritional quality of crop plants under a changing climate: Importance and strategies. *Plant and Soil, 443,* 1–26. https://doi.org/10.1007/s11104-019-04229-0

Tantoh, H. B., Ebhuoma, E. E., Leonard, L. (2022). Indigenous Women's vulnerability to climate change and adaptation strategies in Central Africa: A systematic review. Indigenous knowledge and climate governance: A Sub-Saharan African Perspective, pp 53–66.

UNDESA. (2021). Reconsidering rural development. https://www.un.org/development/desa/dspd/wp-content/uploads/sites/22/2021/05/OVERVIEW_WSR2021.pdf

Vigna, Anne. (2022). Brazil is facing the return of hunger. Le Monde. Rio de Janeiro. Available at: https://www.lemonde.fr/en/economy/article/2022/06/09/brazil-is-facing-the-return-of-hunger_5986229_19.html

von Braun, J., Afsana, K., Fresco, L., Hassan, M., & Torero, M. (2021). Food systems–definition, concept, and application for the UN food systems summit. Sci Innov, 27.

WFP. (2023). *Rome-based agencies.* Available at: https://www.wfp.org/rome-based-agencies

EU and Brazil in the International Circuits of Disavowal of the Climate Crisis

Paula Sandrin

Introduction

Consider the following statement, made by Stefan Agne, Head of Cooperation of the Delegation of the European Union to Brazil at the fourth "Conexão pelo Clima" (Connection for Climate) Fair, held in São Paulo, Brazil, in September 2022, an event that gathered governments, companies, investors, civil society, and stakeholders to discuss opportunities and challenges to the preservation of Brazilian biomes and bioeconomy's implications for a sustainable and socially inclusive economy: "[I]t is perfectly possible to reconcile economic growth and environmental protection by supporting initiatives that work" (in Bressan, 2022, p. 128).

Consider these sentences as well, written by Ignacio Ybáñez, Ambassador of the European Union to Brazil, to a Konrad Adenauer publication: "'Sustainable Connectivity' could be the new guiding principle for EU-Brazil engagement. European Green deal is the bloc's new *growth* strategy, which entails two transitions: green and digital [...] Brazil is a key partner in this agenda. It is one of the world's largest agricultural and food producers. Thanks to the large biodiversity of Brazil's biomes, it has a huge growth potential of its bio-economy [...] Good practice examples and experience in Brazil demonstrate that *it is possible to reconcile economic growth and food production with the protection of the environment*" (Ybáñez, 2022, p. 23, emphasis added).

These sentences are illustrative of a pervasive idea of our current times: it is possible to achieve economic growth while reducing social and environmental impacts and greenhouse gases (GHG) emissions, if political will, economic incentives, natural resources, and technological innovations are made to work in the right direction. The EU would be a perfect example: according to the European Commission (n.d.),

P. Sandrin (✉)
Pontifical Catholic University of Rio de Janeiro, Rio de Janeiro, Brazil
e-mail: paula-sandrin@puc-rio.br

© The Author(s) 2024
A. Ribeiro Hoffmann et al. (eds.), *Climate Change in Regional Perspective*,
United Nations University Series on Regionalism 27,
https://doi.org/10.1007/978-3-031-49329-4_11

169

EU greenhouse gas emissions were reduced by 24% between 1990 and 2019, while the economy grew by around 60% over the same period. The European Green Deal (EGD), approved in 2020, aims to make the EU the first climate-neutral continent by 2050, with an intermediate target of an at least 55% net reduction in GHG emissions by 2030, while still enjoying economic growth.

The EU is not alone. Former UN Secretary General Ban Ki-Moon stressed the need to "decouple economic growth from environmental degradation" as a key element of the Sustainable Development Goals (SDGs) agenda. The Sustainable Development Solutions Network, presided by Jeffrey Sachs, a long-time adviser to the UN Secretary General on SDGs, states that "[t]he key is for all countries, rich and poor, to adopt sustainable technologies and behaviors that decouple economic growth from unsustainable patterns of production and consumption" (Fletcher & Rammelt, 2017, p. 452).

Over the past few years, we have witnessed a global race to announce governmental, financial, and corporate climate ambitions of net-zero targets by 2050 to hold global warming to 1.5 °C above preindustrial times, the more ambitious goal of the Paris Agreement. A net emissions target presupposes that GHG-"positive" emissions can be compensated or cancelled out by "negative" emissions, i.e., removals of GHG from the atmosphere. To achieve net-zero targets, multiple initiatives are advanced: a proper reduction of positive emissions by replacing fossil fuels with renewable energy such as wind, solar, bioenergy, and green hydrogen; compensating emissions through carbon offset markets; removing carbon dioxide through natural or technological "sinks," by restoring damaged ecosystems; protecting tropical forests; and/or developing carbon dioxide removal technologies.

For countries like Brazil, host of the world's largest rainforest with more than 80% of its energy matrix composed of renewables, these targets could bring great opportunities. If restored or preserved, Brazilian forests could act as natural carbon sinks, and the country could be remunerated for providing environmental services, such as carbon storage. According to Joaquim Leite, a former Environmental Minister of far-right Jair Bolsonaro's government (2019–2022) and Head of the Brazilian delegation at COP-26 in Glasgow, the creation of a global carbon market "is fantastic for Brazil, for Latin America, for those who have native forests […] Brazil will be a giant exporter [of carbon credits] in this new green economy, it is a unique opportunity […] We will be a great example to the world. The global market will help and remunerate with credits those who have and take care of native forests. In Brazil, [this are the] cases of the Amazon, Atlantic Forest, Cerrado, Caatinga, and (we can take) care of biodiversity and the community at the same time" (Ministério do Meio Ambiente, 2021). Considering that the carbon market could be bigger than the oil and gas market before 2050 according Forbes, there is a race to turn Brazil into "the Saudi Arabia of the carbon credit" (Ondei, 2021).

More recently, Brazilian President-elect Luis Inácio Lula da Silva announced with great enthusiasm at COP-27 in November 2022 in Egypt an alliance between Brazil, Indonesia, and the Democratic Republic of Congo which calls for a multilateral funding mechanism to help protect half of the world's rainforests, located in the three countries. The forest alliance was nicknamed by British newspaper *The*

Guardian the "OPEC of Rainforests" (Greenfield, 2022). The "paternity" of the forest alliance is disputed between Lula's and Bolsonaro's teams, which evidences how these opposite administrations found "some common ground in pushing for international ecosystem service payment schemes" (Hanbury, 2022).

For EU-Brazil relations, the opportunities are also celebrated. The European Investment Bank invests in solar and wind energy projects in Brazil and aims to convert renewable electricity into green hydrogen to be exported to the EU. Brazil also mines critical raw materials needed for EU's green transition (Ybáñez, 2022, p. 22). The EU Regulation on deforestation-free supply chains, seen in some quarters as a unilateral protectionist measure, could create incentives for Brazil to curb deforestation.

All of this paints a rather rosy picture, which is hard to sustain for several reasons. First of all, there is no scientific evidence supporting the existence of a decoupling of economic growth from environmental impacts "on anywhere near the scale needed to deal with environmental breakdown, but also, and perhaps more importantly, such decoupling appears unlikely to happen in the future" (Parrique et al., 2019, p. 3). Decoupling of economic growth and environmental impact may work on a regional level if polluting activities are outsourced, but there is no empirical evidence that it can work on a global scale. EU's claims that it was possible to achieve economic growth while reducing emissions do not take into account the fact that over the past two decades, imports from China have quadrupled (Pérez, 2021), i.e., polluting activities that impact the environment were outsourced to other regions of the world.

Second, previous offset schemes have largely failed, with emissions reductions from projects often difficult or impossible to measure, short lived, or simply not happening (Friends of the Earth International, 2021). In addition, annual carbon emissions from fossil fuels are greater than the annual amount of carbon that can be captured by natural ecosystems or carbon capture and storage technologies. The carbon dioxide from fossil fuels to be unearthed and burned by 2050 is in addition to the carbon that is already circulating and will remain in the atmosphere for hundreds to thousands of years. Restoring ecosystems – or even planting a trillion new trees – cannot offset continued, additional fossil fuel emissions. Moreover, as the planet's temperature rises, forests and the carbon they store will be increasingly threatened by drought, fires, and soil degradation. Unless there are sinks or drains, to use the vocabulary of the IPCC, the Intergovernmental Panel on Climate Change, large enough to absorb all these emissions, the concentration of CO_2 will continue to increase, as will the temperature of the planet (Friends of the Earth International, 2021).

Technologies that collect or "capture" the carbon dioxide generated by high-emitting activities – such as coal and gas, fired power generation, or plastics manufacturing – are not "drains" large enough either. The 28 carbon capture and storage facilities currently in operation globally have the capacity to capture only 0.1% of fossil fuel emissions and raise concerns about safety and cost of storage (CIEL, 2021). There is, according to the IPCC, "a non-negligible risk of carbon dioxide leakage from geological storage and transport infrastructure" (CIEL, 2021, p. 3).

Finally, renewable energy also depends on raw materials such as lithium, cobalt, and nickel and other rare earth metals whose extraction can have significant environmental and social impacts, thereby once more debunking the idea that economic growth (even with zero GHG emissions) can be decoupled from environmental impacts. In short, decoupling does not work on a global level and currently existing market; technological and nature-based solutions are not enough to mitigate climate change on the scale needed.

All of this is common knowledge. The UN Environment Programme's 2021 Emissions Gap Report (UNEP, 2021), released before COP-26, showed that new national climate pledges combined with other mitigation measures put the world on track to increase global temperatures by 2.7 °C by the end of the century, well above the Paris Agreement goals and generating catastrophic changes in the Earth's climate. To keep global warming below 1.5 °C this century, the world would need to roughly halve annual greenhouse gas emissions by 2030, but they are projected to increase by 13%. The respected Climate Action Tracker updated its predictions after COP-27 and concluded that if all Nationally Determined Contributions (NDCs) for 2030 are met, we will still have a 2.4-degree increase from pre-industrial levels by 2100. The best-case scenario, which assumes the full implementation of all NDCs, all net-zero targets, and all mid-century long-term strategies, predicts a 50/50 chance that warming will be limited to 1.8 °C by 2100 (Climate action Tracker, 2022). In order to stay within the 1.5 °C limit, it would be necessary to urgently promote a "wide-ranging, large-scale, rapid, and systemic transformation" (UNEP, 2022, p. 22). What exactly this means is disputed, but one of possible meaning is turning away from the pursuit of "green" economic growth and replacing the pursuit of efficiency with the pursuit of sufficiency, "that is the direct downscaling of economic production in many sectors and parallel reduction of consumption that together will enable the good life within the planet's ecological limits" (Parrique et al., 2019, p. 3).

This current obsession with net-zero targets, carbon markets, the latest promising renewable energy source, and natural or artificial carbon sinks not only tells us something about the pervasiveness of neoliberal greenwashing in world affairs, about ever-increasing privatization of spheres of life, about an almost religious faith in technological and market solutions, and about the hope that nature will save us and make us profit. This obsession also tells us something about persisting on the wrong path despite all the evidence, which constitutes the mechanism of disavowal. While climate denialism refers to deliberate attempts to deny the reality of the climate crisis, climate disavowal occurs when the climate crisis is recognized, but ineffective responses to deal with it are continually put forward, despite contrary evidence and recurring failures. Disavowal occurs when we admit that something is wrong, that we should be doing something different, but we persist anyway. We recognize the problem, but we deny its significance, its implications, and its gravity, hoping or wishing that things will somehow get solved (with market-based or nature-based or technology-based solutions) despite all the evidence, or because of all the evidence, because that evidence is too painful to bear and the gains of looking away are psychically, symbolic, and materially too great.

In this chapter, I will discuss how this international circuit of disavowal works, which desires and affects animate it, and how the EU and Brazil are caught up in it. In the next section, I will further discuss the concept of climate disavowal, and I will frame decoupling as a fantasy that sustain attachments to capitalism. In the following section, I will illustrate the participation of Brazil and the EU in this international circuit through the case study of a promising new source of "sustainable connectivity" between the two: green hydrogen.

Capitalist Attachments, Climate Disavowal, and Decoupling Fantasy

The international circuit of disavowal of the climate crisis is based on affective attachments to capitalism, understood not only as a mode of accumulation, production, circulation and consumption of capital, goods, and services but as a mode of accumulation and circulation of affective energies, desires, and drives. It is often recognized that capitalism generates a lot of suffering: inequality, expropriation, exploitation, biodiversity loss, pollution, and climate change. In its neoliberal guise, capitalism also generates a lot of psychic suffering: "self-made" entrepreneurial individuals who must invest in themselves, take risks, and be solely responsible for possible "failures," in a context that produces unemployment and precarious jobs. However, it has long been long recognized that capitalism also provides, or at least promises to provide, ample opportunities for the attainment of pleasure. Post-Fordist capitalism does not prevent individuals from having pleasure, nor encourage them postpone it. On the contrary, individuals are commanded to enjoy and are promised various instances of obtaining pleasure, particularly through consumption, which ensures capitalist growth and accumulation (Kapoor, 2020, p. 252). "The naturalization of enjoyment-as-excess […] the desire for the tallest, biggest, wealthiest, flashiest, most original or outrageous […] the ubiquity of the colossal, from beverages to architecture […]; and the surfeit of choice from jeans and fast food to films and dating partners" are evidence that capitalism's productive engine depends on our continuous enjoyment (Kapoor, 2020, p. 103 and 104).

Desires and pleasures are thus not castrated or repressed by the law of God or the state or by the demands of the capitalist economy, but encouraged. Affective energies and desires are not repressed or sublimated, but co-opted by and for capitalist production and marketing. Pleasure and sexuality are incorporated into capitalist culture: the pleasure principle and the reality principle are no longer antagonistic (Brown, 2018). The capitalist regime incites our desires, through the provision of material goods, which are never satisfied, causing our gaze to be cast relentlessly on other objects of desire (Kapoor, 2018). Nowadays, the promise of pleasure is extended to the possibility of self-realization at work, which consequently generates a "passionate servitude" – characteristic of the contemporary configuration of employment (Lordon, 2014). This new configuration is committed to producing joyful affects intrinsic to the work relationship, that is, affects that are not linked to factors external to wage labor such as consumption, for example. Instead, it promises enjoyment inherent to the act of working.

Therefore, in spite of all the suffering it creates, many of us in this world, despite all we know and think, can cling to neoliberal consumerist carbonific capitalism because it structures subjectivities, agencies, ways of life, and desires (for consumption, for social mobility, for economic growth, for development, and for modernization) and promises to satisfy them, even if these promises are never really fulfilled, at least not for everyone. "Green" capitalism also provides or promises to provide abundant opportunities for enjoyment of (electric) personal automobiles, cheap flights (compensating for our carbon footprint) and (energy-efficient) electricity-dependent home appliances, profiting and enjoying while saving the planet. The excitement around the possibility of Brazil becoming "the Saudi Arabia of the carbon credit" and the creation of the "OPEC of the rainforests" not only reveals deeply held attachments to fossil fuels through slips of the tongue but also that enjoyment-as-excess also operates in green capitalism.

If, as individuals, we are interpellated by capitalism as subjects of desire, at a systemic level, capitalism is propelled by an accumulation drive (Kapoor, 2020, p. 76). Drive refers to the compulsion to repeat endlessly, to the continuous circulation around an object in spite of recurrent failures to obtain it. Enjoyment in this case is derived not from obtaining the object (as in desire), but from endless repetition and circulation. In the case of capitalism, drive is manifested as the "endless circular movement of expanded self-reproduction" (Žižek in Kapoor, 2020, p. 79), as "the circular drive to accumulate for the sake of accumulation" (Kapoor, 2020, p. 76). The accumulation drive transform crises into triumphs, as crises (such as the climate crisis) create novel opportunities for accumulation (as in green capitalism). Drive, in this sense, derives pleasure from the challenges of finding new paths to accumulation (Kapoor, 2020, p. 86).

All of this discussion leads to the conclusion that it would be very painful to promote a dramatic transformation of our global socioeconomic system, since it would mean giving up on something that promises pleasure and enjoyment in myriad of forms. In order to avoid the pain that would ensure from breaking attachments to capitalism, we disavow it, i.e., "simultaneously acknowledging and denying our ties (to capitalism) and the pain this causes" (Fletcher, 2018, p. 60). In disavowal, reality – in this case, of climate change – is accepted, but its significance is minimized (Weintrobe, 2013, 7). "[D]isavowal is a simultaneous admission and denial, or a state of 'half-knowing,' that operates according to the formula 'I know very well, but still. . .'" (Žižek in Fletcher, 2018, p. 66). The act of disavowal, instead of completely denying a problem, involves a tendency to quickly fix the problem without out seriously investigating its complex origins and potential solutions (Weintrobe, 2013, p. 8).

In this affective context, mainstream responses to climate change are attractive, appealing, enticing, and difficult to give up on because they promise us that we can, with some adaptations, continue to live, or hope to live, in a way that many of us are invested in. Disavowal, thus, brings great pleasures: the pleasure of avoiding the pain of facing a difficult reality (our efforts are not working, we face a bleak future if we don't give up on something we are strongly attached to); the pleasure of feeling that at least we are doing something: something that is better than nothing, that is not good enough, but that is feasible and "realistic." The international circuit of disavowal is, thus, charged with pleasure.

In this context, decoupling works as a fantasy to mask the real[1] of capitalism and the real of nature, i.e., capitalism is an economic system marked by contradictions that make it unstable unless it continues to expand, and that nature has limits, both in terms of the resources it provides for capitalism and in terms of its ability to offset the effects of capitalism, by capturing carbon, for example (Fletcher, 2018). The fantasy of decoupling hides the underlying conflicts between the goals of reducing inequality, eradicating poverty, ensuring environmental sustainability, reducing GHG emissions, and making profits, which it aims to reconcile. It serves to maintain belief in the idea of achieving SDG's promise of a "sustained, inclusive, and sustainable economic growth" within the framework of a neoliberal capitalist economy (Fletcher & Rammelt, 2017, p 450 and 451).

However, the real insists on rupturing the fantasy. United Nations Environment Programme (UNEP)'s own reports published to inform the SDGs, in spite of optimistic declarations, admits that decoupling is difficult to measure, absolute decoupling is rare, relative decoupling is usually achieved by "a shifting of the material and environmental burden into developing countries" and that "efficiency gains in resource use may paradoxically lead to greater resource use" (UNEP 2011 in Fletcher & Rammelt, 2017, p. 457). UNEP reports indicate that even economic activities that are considered "non-material" have not led to dematerialization. Computerization did not generate a dematerialized "knowledge economy": instead, material extraction, particularly of minerals and ores, increased from 35 billion tons in 1980 to almost 60 billion tons in 2005. This is also recognized by EU's ambassador to Brazil: "Among the common challenges of digitalization faced by both the EU and Brazil are identifying and maximizing synergies between the green and digital agendas. Although digital technologies can help implement climate actions, reduce pollution and restore biodiversity, their widespread use is increasing energy consumption, while also leading to more electronic waste and bigger environmental footprint" (Ybáñez, 2022, p. 20). Other non-material processes that were supposed to support the possibility of decoupling demonstrate similar trends. For instance, the shift from vinyl albums to online music and from books to e-books still requires material production of computers and e-readers, as well as energy to transport these items and power all the equipment used to deliver these digital media (Fletcher & Rammelt, 2017, p. 462). The International Energy Agency forecasts that by 2040 clean-energy technologies will demand 40 times more lithium than in 2020, 25 times more graphite, about 20 times more nickel and cobalt, and 7 times more rare earth elements (The Economist, 2023a).

---

[1] The real can be understood in psychoanalysis as that which manifests as inconsistencies and ruptures in the symbolic order, the order of language and representation.

In spite of all this overwhelming evidence, UNEP and the EU insist on the promises of decoupling. These are clear cases of disavowal: we (UNEP, EU) know very well that there is no evidence that decoupling works (quite the contrary), but still, we are enthusiastic about it.

These attachments to "green" capitalism constitute a relation of cruel optimism: "when something you desire is actually an obstacle to your flourishing" or "actively impedes the aim that brought you to it initially" (Berlant, 2011, p. 1). According to Berlant (2011), this "desired something" can be food, a kind of love, a fantasy of the good life, or a political project. In our case, it is green growth: it draws our attachment by promising to reconcile economic growth, profits, poverty alleviation, environmental protection, and climate change mitigation but actually delivers only (very unevenly distributed) growth and profits, thereby constituting an obstacle to our (humanity, nonhumans, and the planet's) flourishing. As Berlant (2011) explains, in relations of cruel optimism, "the very pleasures of being inside a relation have become sustaining regardless of the content of the relation, such that a person or a world finds itself bound to a situation of profound threat that is, at the same time, profoundly confirming" (p. 2). As we have seen, the abundant evidence that we are heading for a catastrophe unless drastic transformations take place is difficult to bear. Disavowal allows us to avoid this pain while the decoupling fantasy create innumerable opportunities for pleasure. This is how we find ourselves in a situation of profound threat that can be profoundly confirming. Importantly, we should not see cruel optimism as a pathology, perversion or dark truth, but as "a scene of negotiated sustenance that makes life bearable as it presents itself ambivalently, unevenly, incoherently" (Berlant, 2011, p. 14). In a context of ecological collapse, a relation of cruel optimism with green growth makes life bearable at the present time, even if it contributes to unbearable future.

Before we proceed, it is important to note that is not only "green" entrepreneurs, investors, or shareholders, not only policymakers in developed countries and regions (such as the EU), not just consumers in the Global North are part of the international circuit of disavowal. Postcolonial, Third World, Global South countries like Brazil also have aspirations and desires for growth, modernization, and development (Chakrabarty, 2018). In fact, even some of the losers of neoliberal capitalism (the indebted, those in precarious jobs in the South and the Global North), even some of the staunchest critics of capitalism, can be caught up in circuits of disavowal. Thus, it is not only the global political, financial, and industrial elites that refuse to promote the drastic socioeconomic transformations needed to mitigate the climate crisis but also the middle and subaltern classes of Asia, Africa, and Latin America who cling to the promises of capitalism and modernization.

The Cruel Optimism of Green Hydrogen and the Phantasmatic Promise of Sustainable Connectivity Between EU and Brazil

In this section, I will explore EU-Brazil cooperation on clean energy, particularly the promises of green hydrogen, which is currently being presented as one of the most promising sources of "sustainable connectivity" between EU and Brazil. I will argue the current excitement with green hydrogen follows the affective structure of a cruel optimistic attachment: it literally and metaphorically adds fuel to the fantasy that we can continue to pursue economic growth without facing catastrophic consequences, because, in this case, an alliance between nature and technology will save us.

The protection of the environment and cooperation in energy, specifically sustainable biofuels, have been a focal point of the EU-Brazil strategic partnership since 2007. The European Green Deal and the invasion of Ukraine by Russia have further strengthened EU's goal of reducing its dependence on fossil fuels. EU's Joint Action for More Sustainable, Affordable and Secure Energy (REPowerEU) aims to decrease Europe's dependence on Russian oil and gas by 2030 by diversifying its imports and relying more on renewable energy sources in the long run. The target is to meet 45% of the EU's energy needs through renewable energy sources such as sustainably produced bioenergy (i.e., bioenergy that does not involve deforestation or food insecurity), solar and wind power, and green hydrogen generated from renewable sources (Boehm, 2022, p. 39).

As countries present their net-zero plans to meet global climate goals, hydrogen has regained importance on the agendas of the EU as well as countries like Australia, the UK, and Japan, who have released national hydrogen strategies. In July 2020, the European Commission released its "Hydrogen Strategy for a climate-neutral Europe," which included a bold objective of achieving 40 gigawatts (GW) of electrolyzer capacity by 2030 for the production of green hydrogen (for comparison, between 2000 and 2019, a total capacity of just 0.25 GW of green hydrogen projects was deployed globally) (Carbon Brief, 2020). European Commission vice-president Frans Timmermans said that "[Hydrogen] has become the rockstar of new energies all across the world"; the International Energy Agency (IEA), the Hydrogen Council, Shell and BP have also shared their visions for the future importance of hydrogen; and British Newspaper *The Telegraph* has dubbed the 2020s as the "hydrogen decade" (Carbon Brief, 2020).

This is because hydrogen burns cleanly, releasing only water and energy; contains more energy per unit of weight than traditional fossil fuels; and also allows energy to be stored and transported (Thomaz & Pimentel, 2022, p. 62). In a positive scenario, hydrogen may have the ability to fuel transportation modes such as trucks, planes, and ships. Additionally, it could serve as a heating source for homes, stabilize electricity grids, and aid heavy industry in producing materials like steel and cement. Hydrogen could be particularly useful to sectors which are difficult to electrify, such as steel production, shipping, and aviation (Carbon Brief, 2020). However, achieving all these functions with hydrogen would necessitate vast amounts of the fuel, and its cleanliness would be dependent on the methods utilized to create it. Hydrogen is the most abundant element on Earth, but it doesn't exist in its pure form, so it has to be separated for use in energy production. There are various ways to extract hydrogen, each varying in terms of their level of renewability and cleanliness, as well as the technology employed. Grey hydrogen is made from fossil fuels; blue hydrogen is also made from fossil fuels, but with capture and storage of carbon dioxide emissions; green hydrogen is generated using electrolysis powered by renewable electricity and does not emit GHG gases (Carbon Brief, 2020).

International trade in green hydrogen is growing between producing countries like Brazil and other countries in Latin America, Africa, Australia, and the Caribbean and importing countries primarily in Europe and Asia, who do not have enough renewable energy sources to achieve their goal of decarbonization (Panik, 2022, p. 96). Although green hydrogen production from renewable energy sources is still expensive, countries with abundant resources could have the potential to produce it at lower costs. Brazil operates onshore wind projects with some of the highest capacity factors in the world. According to the Bloomberg "Hydrogen Economy Outlook" report, Brazil has the second-lowest cost of producing green hydrogen after China (Panik, 2022, p. 95). Brazil's government has created a National Hydrogen Program aimed at developing its hydrogen market and has also set up a credit line and is investing in research, development, and innovation to support its hydrogen initiatives (Thomaz & Pimentel, 2022, p. 63).

The State of Ceará in Brazil launched its first Green Hydrogen Hub in February 2021 and has since signed over 20 Memoranda of Understanding with national and international companies that have announced investments worth around USD 20 billion in the Pecém Green Hydrogen Hub (Panik, 2022, p. 97). Other states in Brazil have also announced plans to launch their own Green Hydrogen Hubs. The European Investment Bank has invested €1 billion in solar and wind energy projects in Brazil and aims to convert this renewable energy into green hydrogen for export to Europe (Ybáñez, 2022). Three major infrastructure projects are being planned in Ceará, Rio de Janeiro, and Pernambuco for the export of green hydrogen to Europe, and the Port of Rotterdam owns 30% of the Port of Pecém (Panik, 2022, p. 97).

All of this again paints a rather rosy picture that masks the real of green hydrogen. The IEA's, 2019 hydrogen report says that "it may be tempting to envisage an all-encompassing low-carbon hydrogen economy in the future," but it adds that "the use of hydrogen in certain end-use sectors faces technical and economic challenges compared with other (low-carbon) competitors" (IEA, 2019). Michael Liebreich, senior contributor to BloombergNEF, wrote that "On the surface, [hydrogen] seems like the answer to every energy question" but added that "Sadly, hydrogen displays an equally impressive list of disadvantages" (Liebreich, 2020). The list is quite long, but it can be summarized as follows.

First, the majority of hydrogen production currently comes from high-carbon sources without carbon capture, which is cheaper than using renewable sources or capture and storage technologies. The IEA reports that the majority (76%) of hydrogen today is produced from gas and 23% from coal, mostly in China, while only 2%

is produced through electrolysis (IEA, 2019, p. 37). Second, hydrogen is challenging to transport and store due to its bulkiness and cost. While importing hydrogen from sunny or windy regions, such as Brazil, may be an attractive option for countries lacking sufficient renewable resources, the cost of transporting hydrogen in special containers at high pressures and low temperatures is expensive. The hydrogen industry is currently very localized, with 85% of hydrogen being produced and used on-site due in part to the high costs of transporting it (IEA, 2019, p. 68). Additionally, distributing hydrogen within a country requires significant upfront investment in infrastructure. Third, its production and use are also not very efficient when compared to alternative energy sources, leading to increased expenses and energy requirements. Electric vehicles, for example, are several times more efficient than hydrogen fuel cell vehicles (Carbon Brief, 2020). Furthermore, the supply and value chains involved in the use of hydrogen are complex, requiring coordination among various parties.

The IEA acknowledges the uncertainty regarding the costs of producing hydrogen from various sources in different regions, as well as how they will compete in the future. However, the IEA does note that the cost of green hydrogen could decrease by 30% by 2030 due to the declining costs of renewables and the scaling up of hydrogen production (IEA, 2019, p. 14). It's worth mentioning, however, that comparing studies is challenging since some studies only include production costs, while others include transport and distribution costs. Ultimately, fossil fuel costs will have the most significant impact on future hydrogen prices, and the success of green and blue hydrogen will depend on future electricity and gas costs (Carbon Brief, 2020).

In addition, hydrogen may have a significant impact on global trade, potentially creating a new group of energy exporters rich in solar and wind energy, such as Chile, Australia, and Morocco. This development could also potentially reshape geopolitical relations and alliances between countries. The Economist has even coined the term "electrostates" to describe these future energy powerhouses. However, there are concerns about the potential for "green colonialism" in the hydrogen revolution, as developing countries could be viewed solely as providers of raw materials (Van de Graaf et al., 2020).

Even if we leave the important issues of green colonialism, the potential for dematerialization, and environmental impacts aside, it is clear that creating a market for green hydrogen that has the necessary infrastructure and is competitive with other energy sources is challenging. To make green hydrogen competitive, the "ideal conditions" of low renewable energy costs, low investment costs, and high operating hours must come together. This will require policies to be put in place within the green hydrogen production process to reduce costs, increase capacity, and promote domestic electrolysis production through incentives such as tax cuts, subsidies, loan credits, feed-in tariffs, and others (Thomaz & Pimentel, 2022, p. 68). Brazil has the potential to become a green hydrogen exporter *if* it can produce green hydrogen at a competitive price while maintaining sustainability standards and having a transportation-friendly grid (Thomaz & Pimentel, 2022, p. 72).

As we can see, there are a lot of "ifs" and conditions to be met. This is, in fact, a common linguistic structure of optimistic proclamations of the possibilities of green growth, which usually begins with "(a) a statement of a win-win scenario," i.e., the opportunities of green hydrogen, and is followed by "(b) a caveat in the form of a reality check usually starting with a 'but', which emphasizes the challenges in achieving the desired win-win scenario" (Oya in Fletcher & Rammelt, 2017, p. 456). The difficulties, in this case, reside not only in amassing political will, achieving the right market prices and developing technological innovations.

The current excitement with green hydrogen follows the affective structure of a cruel optimistic attachment: a "sustaining inclination to return to the scene of fantasy [in this case, again that it is possible to decouple economic growth from GHG emissions] that enables you [the EU, Brazil] to expect that this time, nearness to this thing [green hydrogen] will help you or a world to become different [*really* mitigating climate change] in just the right way" (Berlant, 2011, p. 2). This optimism is cruel because the object that draws our attachment, decoupling, the latest clean energy source, green hydrogen, actively impedes the aim – mitigating climate change – that brought us to it in the first place. Green hydrogen hampers the goal of mitigating climate change because it diverts us from actually downscaling economic production and consumption. It literally and metaphorically adds fuel to the fantasy that we can continue to pursue economic growth without facing catastrophic consequences, because, in this case, the alliance between technology and nature (the sun, wind, and hydrogen as the most abundant substance on Earth) will save us.

Green hydrogen is not the panacea that will deliver growth decoupled from environmental impacts, but the excitement around it suggests that it is working in other spheres. It generates new investments opportunities; pleasure from finding new paths to accumulation; activities in which one can feel entrepreneurial, proactive, useful, creative, and virtuous; and hope for countries Brazil to become an "electrostate" (as well as the "Saudi Arabia of Carbon Credits" and member of the "OPEC of the rainforests"). Green hydrogen sets in motion an array of activities that deliver abundant material and affective gratifications.

Concluding Remarks

The Economist (2023b), in a recent Technology Quarterly, argued that one of the obstacles to greening electricity grids, paramount to mitigate climate change, lie in some environmentalists who oppose building the necessary big bits of infrastructure: "The skyline must be preserved, they might say, or the woodland is too ancient to fell, or the colony of terns too important in and of itself." According to the magazine, "more building is the most practical course of action […] and it is economic growth that will make possible the building of new transmission lines, gigawatt-scale renewable power installations and, indeed, the mines from which the minerals these things need are sourced. To demonise it, as some environmentalists do, is to expose the world to more climate change, not less." It warns "those who believe there is no way to stop climate change through growth" that "to change the way the world thinks, person by person, is a yet more ambitious task than changing the ways in which the world generates and distributes its electric power. If the energy transition cannot be achieved with the habits of mind already available, it is hard to see that it can be achieved at all."

Yet again we see an example of disavowal at work, propelled by attachment to capitalist economic growth. The magazine knows that it is impossible to decouple green electricity grids from environmental impact: it sarcastically notes that the skyline, the woodland, and the colony of terns will suffer; minerals will need to be mined. And yet, this course of action is defended because it is the most practical, the only feasible one to mitigate climate change. Questioning the continuous pursuit of economic growth, or the "GOD imperative" (Grow or Die) (Kapoor, 2020, p. 80) upon which capital circulation depends, is ruled out as impractical.

This chapter called attention to the affective energies, desires, and drives that sustain the international circuit of disavowal of the climate crisis, predicated on attachment to capitalism and sustained by the fantasy of decoupling. As we have seen, all the available scientific evidence shows that the solutions that have been continually put forward by the international climate regime, including green electricity grids and green hydrogen, cannot work fast enough or on the scale needed to prevent the Earth's climate from exceeding 1.5 °C above preindustrial levels, to say nothing of the non-negligible environmental and social impacts. These solutions, thus, are not practical or realistic, but they surely are enjoyable. This chapter departed from the premise that identifying and making sense of the circuits of pleasures and displeasures that underlie and are set in motion by ecological collapse is crucial if we are to make an accurate diagnosis of the present, which is a condition for imagining and practicing really effective responses.

Although focused on Brazil-EU relations, the mechanisms of disavowal described here also apply to the relations between the EU and Latin America more generally. After all, Latin American countries are full of resources needed to achieve net-zero targets and carry out the green transition: tropical forests, renewable energy sources, and critical minerals such as lithium and copper and rare earth elements are just some of them. The risk of "green colonialism" and hazardous environmental impacts of mining are not negligible, but they are usually sidelined by (cruel) optimist discourses that present Latin America as key to the global energy transition. By focusing on the affective attachments that stand in the way of effective policy responses to this global problem, this chapter hopefully contributes to discussions on interregional cooperation on climate change as well.

References

Berlant, L. (2011). *Cruel optimism*. Duke University Press.
Boehm, L. M. (2022). Global implications of Europe's energy crisis. In R. J. Themoteo (Ed.), *Cooperation between Brazil and Europe: geopolitical importance and innovation perspectives*. Konrad Adenauer Stiftung, Série Relações Brasil-Europa 12, 2022. Available at: https://www.kas.de/documents/265553/265602/SRBE+12+web.pdf/fb1a8603-431a-c507-fc4a-e57dbb48 a9ee?version=1.0&t=1673628310948. Accessed on February 07, 2023.

Bressan, R. N. (2022). Bioeconomy: potential cooperation between Brazil and Europe. In: R. J. Themoteo (Ed.), *Cooperation between Brazil and Europe: Geopolitical importance and innovation perspectives*. Konrad Adenauer Stiftung, Série Relações Brasil-Europa 12, 2022. Available at: https://www.kas.de/documents/265553/265602/SRBE+12+web.pdf/fb1a8603-431a-c507-fc4a-e57dbb48a9ee?version=1.0&t=1673628310948. Accessed on February 07, 2023.

Brown, W. (2018). Neoliberalism's Frankenstein: Authoritarian freedom in twenty-first century "democracies". *Critical Times, 1*(1), 60–79.

Carbon Brief. (2020). In-depth Q&A: Does the world need hydrogen to solve climate change? 30/09/2020. Available at: https://www.carbonbrief.org/in-depth-qa-does-the-world-need-hydrogen-to-solve-climate-change/ Accessed on February 07, 2023.

Chakrabarty, D. (2018). Planetary crisis and the difficulty of being modern. *Millennium: Journal of International Studies, 46*(3), 259–282.

CIEL. (2021). Confronting the myth of carbon-free fossil fuels: Why carbon capture is not a climate solution. *Center for International Environmental law*, July 2021. Available at: https://www.ciel.org/wp-content/uploads/2021/07/Confronting-the-Myth-of-Carbon-Free-Fossil-Fuels.pdf. Accessed on February 07, 2023.

Climate Action Tracker. (2022). The CAT thermometer. Available at: https://climateactiontracker.org/global/cat-thermometer/. Accessed on February 07, 2023.

European Commission. (n.d.). Progress made in cutting emissions. Available at: https://climate.ec.europa.eu/eu-action/climate-strategies-targets/progress-made-cutting-emissions_en. Accessed on February 07, 2023.

Fletcher, R. (2018). Beyond the end of the world: Breaking attachment to a dying planet. In I. Kapoor (Ed.), *Psychoanalysis and the GlObal*. University of Nebraska Press.

Fletcher, R., & Rammelt, C. (2017). Decoupling: A key fantasy of the Post-2015 sustainable development agenda. *Globalizations, 14*(3), 450–467. https://doi.org/10.1080/1474773 1.2016.1263077

Friends of the Earth International. (2021). Chasing carbon Unicorns: The deception of carbon markets and "net zero". Available at: https://www.foei.org/wp-content/uploads/2021/04/Friends-of-the-earth-international-carbon-unicorns-english.pdf. Accessed on February 07, 2023.

Greenfield, P. (2022). Brazil, Indonesia and DRC in talks to form 'Opec of rainforests'. *The Guardian*, 05/11/2022. Available at: https://www.theguardian.com/environment/2022/nov/05/Brazil-Indonesia-drc-cop27-conservation-opec-rainforests-aoe. Accessed on February 07, 2023.

Hanbury, S. (2022). Where is the money? Brazil, Indonesia and Congo join forces in push for rainforest protection cash. *Mongabay*, 18/11/2022. Available at: https://news.mongabay.com/2022/11/where-is-the-money-Brazil-Indonesia-and-Congo-join-forces-in-push-for-rainforest-protection-cash/. Accessed on February 07, 2023.

IEA. (2019). The Future of Hydrogen, Seizing today's opportunities. Report prepared by the IEA for the G20, Japan. June 2019. Available at: https://iea.blob.core.windows.net/assets/9e3a3493-b9a6-4b7d-b499-7ca48e357561/The_Future_of_Hydrogen.pdf. Accessed on April 07, 2023.

Kapoor, I. (2020). *Confronting desire - psychoanalysis and international development*. Cornell University Press.

Kapoor, I. (2018). *Psychoanalysis and the GlObal*. University of Nebraska Press.

Liebreich, M.(2020). Separating hype from hydrogen – Part one: The supply side. *BloombergNEF*, 08/10/2020. Available at: https://about.bnef.com/blog/liebreich-separating-hype-from-hydrogen-part-one-the-supply-side/ Accessed on April 07, 2023.

Lordon, F. (2014). *Willing slaves of capital: Spinoza and Marx on desire*. Verso.

Ministério do Meio Ambiente. (2021). Brasil liderou negociações que resultaram na criação do mercado global de carbono. Available at: https://www.gov.br/mma/pt-br/noticias/brasil-liderou-negociacao-para-criacao-do-mercado-global-de-carbono. Accessed on February 07, 2023.

Ondei, V. (2021). Mercado de carbono pode ser maior que o de óleo e gás, hoje em US$ 2 trilhões, antes de 2050. *Forbes*, 25/08/2021. Available at: https://forbes.com.br/forbesesg/2021/08/mercado-de-carbono-pode-ser-maior-que-o-de-oleo-e-gas-hoje-em-us-2-trilhoes-antes-de-2050/. Accessed on February 07, 2023.

Panik, M. S. (2022). Energy transition in Brazil and the European Union: developments and potential for cooperation in the field of clean energy. In R. J. Themoteo (Ed.), *Cooperation between Brazil and Europe: geopolitical importance and innovation perspectives*. Konrad Adenauer Stiftung, Série Relações Brasil-Europa 12, 2022. Available at: https://www.kas.de/documents/265553/265602/SRBE+12+web.pdf/fb1a8603-431a-c507-fc4ae57dbb48a9ee?version=1.0&t=1673628310948. Accessed on: 07/02/2023.

Parrique, T., Barth, J., Briens, F., Kerschner, C., Kraus-Polk, A., Kuokkanen, A., & Spangenberg, J. H. (2019). *Decoupling debunked: Evidence and arguments against green growth as a sole strategy for sustainability*. European Environmental Bureau.

Pérez, A. (2021). A Green New Deal for whom? *Open Democracy*, 23/04/2021. Available at: https://www.opendemocracy.net/en/oureconomy/green-new-deal-whom/. Accessed on February 07, 2023.

The Economist. (2023a). How America plans to break China's grip on African mineral. A new contest between the US and China is under way. *The Economist*, 28/02/2023. Available at: https://www.economist.com/middle-east-and-africa/2023/02/28/how-america-plans-to-break-chinas-grip-on-african-minerals?utm_content=article-link-2&etear=nl_today_2&utm_campaign=a.the-economist-today&utm_medium=email.internal-newsletter.np&utm_source=salesforce-marketing-cloud&utm_term=2/28/2023&utm_id=1504889. Assessed on March 01, 2023.

The Economist. (2023b). The case for an environmentalism that builds. *The Economist*, 05/04/2023. Available at: https://www.economist.com/leaders/2023/04/05/the-case-for-an-environmentalism-that-builds. Assessed on March 07, 2023.

Thomaz, L. Forti & Pimentel, N. F. (2022). The potential of bioenergy and green hydrogen for the Brazil-European Union cooperation. In R. J. Themoteo (Ed.), *Cooperation between Brazil and Europe: geopolitical importance and innovation perspectives*. Konrad Adenauer Stiftung, Série Relações Brasil-Europa 12, 2022. Available at: https://www.kas.de/documents/265553/265602/SRBE+12+web.pdf/fb1a8603-431a-c507-fc4a-e57dbb48a9ee?version=1.0&t=1673628310948. Accessed on February 07, 2023.

UNEP. (2021). *The heat is on. A world of climate promises not yet delivered*. Emissions gap report 2021. Available at: https://www.unep.org/resources/emissions-gap-report-2021. Accessed on February 10, 2023.

UNEP. (2022). *The closing window climate crisis calls for rapid transformation of societies*. Emissions gap report 2022. Available at: https://www.unep.org/resources/emissions-gap-report-2022. Accessed on February 10, 2023.

Van de Graaf, T., et al. (2020). The new oil? The geopolitics and international governance of hydrogen. *Energy Research & Social Science, 70*, 101667. Published online June 30, 2020. https://doi.org/10.1016/j.erss.2020.101667

Ybáñez, I. (2022). EU and Brazil – sustainable connectivity. In: R. J. Themoteo (Ed.), *Cooperation between Brazil and Europe: geopolitical importance and innovation perspectives*. Konrad Adenauer Stiftung, Série Relações Brasil-Europa 12, 2022. Available at: https://www.kas.de/documents/265553/265602/SRBE+12+web.pdf/fb1a8603-431a-c507-fc4a-e57dbb48a9ee?version=1.0&t=1673628310948. Accessed on February 07, 2023.

Weintrobe, S. (2013). Introduction. In S. Weintrobe (Ed.), *Engaging with climate change: Psychoanalytic and interdisciplinary perspectives* (pp. 1–15). Routledge.

Conclusions

Andrea Ribeiro Hoffmann, Paula Sandrin, and Yannis E. Doukas

The environment and climate change are addressed in a myriad of multilateral, regional, and bilateral channels of dialogue, negotiations, and cooperation mechanisms; our focus in this volume was on the regional level. We set the aim to reflect critically on ongoing agendas on the environment and climate change in Europe and Latin America and to contribute to the dialogue between the European Union (EU) and the Community of Latin American and Caribbean States (CELAC) by advancing recommendations on how to improve the mechanisms to address these issues in the context of the EU-CELAC bi-regional strategic partnership. While in Europe the EU is a regional actor and a focal point for the debate, formulation, and implementation of environmental and climate change initiatives, in Latin America, there are many regional organizations; CELAC is the largest in membership, encompassing all 33 LAC countries, but it is primarily a political forum for debate and articulation of consensus; it does not formulate or implement policies. For this reason, we included in our analyses other regional organizations and institutions, such as MERCOSUR, UNASUR, PROSUR, ECLAC, regional development banks, and regional civil society alliances, and their relations with global multilateral institutions such as the United Nations System.

This edited volume was structured in three main sections: a section discussing the norms, institutions, and agenda on the environment and climate change within and among the EU and Latin American regional organizations; a section addressing the challenges to finance development and a "greener" economy; and a section assessing so-called new green solutions to climate change, including in the

A. Ribeiro Hoffmann (✉) · P. Sandrin
Pontitical Catholic University of Rio de Janeiro, Rio de Janeiro, Brazil
e-mail: a_ribeiro_hoffmann@puc-rio.br; paula-sandrin@puc-rio.br

Y. E. Doukas
National and Kapodistrian University of Athens, Athens, Greece
e-mail: jodoukas@pspa.uoa.gr

© The Author(s) 2024
A. Ribeiro Hoffmann et al. (eds.), *Climate Change in Regional Perspective*,
United Nations University Series on Regionalism 27,
https://doi.org/10.1007/978-3-031-49329-4_12

agriculture sector. The chapters contribute to the understanding of to what extent and how the regional level contributes to policy making in the fields of environment and climate change, vis-à-vis the global and the national levels by addressing various aspects of the four key axes of regional policy making advanced in the Introduction – regional redistribution mechanisms, regulations, rights, and cooperation.

From a comparative regionalism perspective, it is possible to affirm that in the fields of environment and climate change, the discrepancy between the EU and Latin American regional organizations is significant. EU policies are not a panacea, but a lot has been discussed and done to address the environment and climate change. Most authors have praised initiatives such as EU Green Deal (2019), the EU Climate Law (2021), EIB green mechanisms, and CAPs Agenda 2000, despite concerns such as procrastination, effectiveness, and asymmetric power relations among urban consumers and rural farmers.

Latin American countries have increasingly addressed environmental concerns and climate change in their domestic policies and in their participation in global multilateral institutions over the last years, but Latin American regional organizations have felt short of including these matters in their priorities. It is open to debate whether the regional level of social policy making may make a difference in Latin America, but as long as regional organizations include economic commitments, more or less liberal, orthodox, or heterodox, it is imperative that considerations about the environment and climate change are taken seriously, to at least avoid negative (non-intended) effects.

The remaining of this concluding chapter presents a compilation of key recommendations advanced by the authors of this edited volume to address the climate change. Given the interdisciplinary nature of the volume, which includes theoretical perspectives from international relations, law, economics, global ethics, and psychoanalysis, the recommendations refer both to the political and the policy level, as well as the institutional design of cooperation.

Diz & Oliveira (2024) explore the innovations and potential effectiveness and impact of the European Green Deal and the consequent European Climate Law in the local, regional, national, intra-community, and international relations of the EU, from a legal perspective and an analytical-conceptual and dogmatic-propositional methodology. They argue that the EU is a leader in proposing mechanisms such as policies, programs, and actions aiming to provide a fair, efficient transition that encompasses all productive sectors, especially those with the greatest impact on GHG emissions, but that the current context of economic and political crisis, generated by the pandemic and the war in Ukraine has, however, destabilized some of the efforts made so far. They recommend States and the EU to continue to make joint efforts to implement regulations that will minimize the effects caused by climate change and policies aimed at climate neutrality, including in the EU-Mercosur agreement.

Ribeiro Hoffmann (2024) draws on historical institutionalism and the concept of epistemic communities to assess the role of Latin American regional organizations in addressing climate change. She argues that despite the activism of these organizations during the last decades, they have not addressed climate change as a priority and that had (unintended) negative effects on the environment given their approach to development. The lack of epistemic communities and the strong lobby of agrobusiness sectors in most LAC countries, including Brazil, hindered the incorporation of stronger agendas at the regional level. Her main recommendation is that LAC regional organizations foment debates and include commitments to address environmental and climate change problems as a priority. She argues that the current context and critical juncture may provide a space for new initiatives at the regional and global level, including the EU-CELAC partnership.

Castiglioni (2024) argues that a key problem of the bi-regional relations is that each region departs from different assumptions about the nature of the problem of climate change: while the EU tries to reconcile commercial and political interests with the aim to act as a normative power in global governance, regions from the Global South such as MERCOSUR criticize EU's attempts to ensure binding commitments in bi-regional agreements as imperialist or a cover trade interests. He explains these patterns drawing on the literature of global ethics and the debate between universal and virtue ethics, and universal and communitarian values, but he suggests a path to overcome this dichotomy based on the concept of the environment as a "global common good." In this context, the allocation of responsibilities to different actors, including corporate social responsibility, might provide a path for the regions to overcome their differences.

Ghymers (2024) develops an analysis based on economic systemic theory and identifies the underpricing of fossil energies as the key problem for tackling climate change. Short-term political interests and vested economic interests lead to "irrational procrastination." He proposes to scrap subsidies to fossil energies, foster clean alternatives, and transfer resources to low-income households. To achieve this, he recommends the implementation of the "two-steps-two-tier method" to create trust among peers and foster the creation of collective mechanisms for monitoring the energy transition. The two-tier scheme in which experts dialogue in consultive, non-decision fora under confidential rules at the regional level is reproduced at the bi-regional level. He argues that while the EU and LAC cannot solve climate change problems on their own, a consensus among them would provide an important path to global-level cooperation.

Griffith-Jones & Carreras (2024) discuss the role of developing banks and more specifically the European Investment Bank to finance the green transformation. They argue that these banks' tools such as the use of carbon shadow pricing to evaluate projects and venture debts are powerful mechanisms to allocate resources. They argue that the EIB has been central to financing the European Green Deal, it has stopped funding fossil fuels, and by 2025, it will have 50% of its lending in climate change-related activities, both mitigation and adaptation. They recommend the bank to continue this path of action and expand its funding abroad, including to Latin America.

Schulmeister (2024) draws on economic theory and theories of collective action and proposes an alternative approach to conventional carbon pricing and emission trading schemes, namely, the determination of a path of steadily rising prices in fossil energies such as oil, coal, and natural gas. The rising prices are to be accompanied by a monthly adjusted quantity tax to match the difference with global market prices, and these resources from fossil energy taxes can be used to fund ecological transformation. Since there is no world state to set these taxes, regional organizations such as the EU could lead the initiative. He argues that this is a better alternative to degrowth strategies, which in his view are not good as economic activities should aim at providing the basis for a good life for the greatest possible number of people, and the problem we currently face is more of collective action than economic.

Doukas, Vardopoulos & Petides (2024) draw on historical institutionalism to analyze the gradual integration of agri-environmental measures in the European Common Agriculture Policy's (CAP) policy making, during the last two decades. They pay attention to the fact that these changes affect power relations among the stakeholders involved and recommend that the measures address the concerns of the European farmers and balance their needs with the environment, and the concerns of the society. In this sense, measures such as payments for ecosystem services and the use of target and flexible policies based on farm and local context can be promising to cope with the sustainability challenge. They also recommend that socioeconomic disparities in rural areas are addressed and that policies should avoid the concentration of land ownership and displacement of small farmers.

Maravegias, Doukas & Petides (2024) discuss the potential effects of the changes introduced by CAP in the area of research and innovation (R&I), in particular, the adoption of digital technologies and data to support the transition to a climate-neutral, circular, and resilient economy. They argue that technology applications are carried out locally, but their production is highly internationalized and concentrated in private companies, mostly multinational. These applications also require farmers to have professional training and support the initiatives of the European Agricultural Fund for Rural Development in this regard. Despite this, the authors call for further discussion on the implications of such transformations, and the potential economic disparities between the participants in the global food systems, within in the EU, and worldwide. This matter is particularly relevant to EU-CELAC relations given the place of agriculture in the bi-regional relations.

Teixeira & Jardim (2024) discuss the links between climate change, food systems, and food and nutrition security from the perceive of a human rights-based approach to development and the Food Nutrition Security (FNS) approach. They advocate that the right to adequate food must be placed at the center of the strategies to mitigate and adapt to the impacts of climate change and five recommendations: promote multi-stakeholder partnerships and an intersectoral approach; foster nutrition-sensitive and climate-resilient value chains; address finance gaps; include gender equality and women's empowerment perspective; and include a social behavior change communication strategy. They illustrate the potential of their proposals in a case study, IFAD's Pro-Semiarid Project in Bahia – Brazil; this example and the best practices they analyze can be used in the EU-CELAC cooperation.

Sandrin (2024) unveils flaws of current proposals to mitigate climate change based on the concept of disavowal from a psychoanalytical perspective. Unlike climate denial, actors in circuits of climate disavowal do acknowledge climate change but insist on implementing ineffective responses notwithstanding ample contrary evidence and recurring failures. She illustrates this process with a case of EU-Brazil cooperation and the most recent project on green hydrogen, which she assesses as unlikely to contribute to tackling climate change and argues that this process creates (enjoyable) illusions that something is being done but recommends that a more accurate diagnosis of the current situation is done in order to avoid ecological collapse.

Despite the variation in theoretical perspectives, proposed mechanisms, and level of optimism, the authors of his edited volume agree on the pressing need to address the environment and climate change. Regional organizations in Europe and Latin America are relevant spaces for the debate, formulation, and implementation of redistribution mechanisms, regulations, rights, and cooperation in these fields and have a potential to complement the national and global level efforts. Further research, discussion, and actions are needed to avoid an ecological tragedy, and we hope that this volume is a small contribution in the direction to prioritize that agenda!

References

Castiglioni, F. (2024). An "aggressive" cooperation: Environment as a hot issue in EU-LAC relations. In A. R. Hoffmann, P. Sandrin, & Y. E. Doukas (Eds.), *Climate change in regional perspective*. Springer Nature Switzerland AG.

Diz, J. B. M., & de Oliveira, M. L. (2024). The EU in a multidimensional regime: The regulation of climate neutrality. In A. R. Hoffmann, P. Sandrin, & Y. E. Doukas (Eds.), *Climate change in regional perspective*. Springer.

Doukas, Y. E., Vardopoulos, I., & Petides, P. (2024). Challenging the status quo: A critical analysis of the common agricultural policy's shift toward sustainability. In A. R. Hoffmann, P. Sandrin, & Y. E. Doukas (Eds.), *Climate change in regional perspective*. Springer

Ghymers, C. (2024). Fostering the dynamics of the Bi-regional Summit EU-CELAC for spurring the cooperation in climate change. In A. R. Hoffmann, P. Sandrin, & Y. E. Doukas (Eds.), *Climate change in regional perspective*. Springer.

Griffith-Jones, S., & Carreras, M. (2024). The role of European Investment Bank (EIB) and national and regional development banks in the green transformation. In A. R. Hoffmann, P. Sandrin, & Y. E. Doukas (Eds.), *Climate change in regional perspective*. Springer.

Maravegias, N., Doukas, Y. E., & Petides, P. (2024). Climate change concerns and the role of research and innovation in the agricultural sector: The European union context. In A. R. Hoffmann, P. Sandrin, & Y. E. Doukas (Eds.), *Climate change in regional perspective*. Springer.

Ribeiro Hoffmann. (2024). Climate change coopertation in LAtin American regionalism. In A. R. Hoffmann, P. Sandrin, & Y. E. Doukas (Eds.), Climate change in regional perspective. Springer.

Sandrin, P. (2024). EU and Brazil in the international circuits of disavowal of the climate crisis. In A. R. Hoffmann, P. Sandrin, & Y. E. Doukas (Eds.), *Climate change in regional perspective*. Springer.

Schulmeister, S. (2024). Fixing rising price paths for fossil energy: Basis of a "green growth" without rebound Effects. In A. R. Hoffmann, P. Sandrin, & Y. E. Doukas (Eds.), *Climate change in regional perspective*. Springer.

Teixeira, A., & Jardim, C. A. (2024). Building climate-resilient food systems: The case of IFAD in Brazil's Semiarid. In A. R. Hoffmann, P. Sandrin, & Y. E. Doukas (Eds.), *Climate change in regional perspective*. Springer.